Energy Processing in Cells
Recurring Concepts and Themes

Energy Processing in Cells
Recurring Concepts and Themes

Michael Guppy

ATHENA PRESS
LONDON

Energy Processing in Cells
Recurring Concepts and Themes
Copyright © Michael Guppy 2008

All Rights Reserved

No part of this book may be reproduced in any form,
by photocopying or by any electronic or mechanical means,
including information storage or retrieval systems,
without permission in writing from both the copyright owner
and the publisher of this book.

ISBN 10-digit: 1 84401 676 5
ISBN 13-digit: 978 1 84401 676 1

First Published 2007 by
ATHENA PRESS
Queen's House, 2 Holly Road
Twickenham TW1 4EG
United Kingdom

Printed for Athena Press

Acknowledgements

This book arose from a lecture series on metabolism that was developed, refined and clarified over twenty years. The bulk of the book was written while I was a staff member at the University of Western Australia.

I dedicate this book to the science and medical undergraduate students at the University of Western Australia between 1985 and 2003. The curiosity of these students demanded that I understood and then clarified the concepts involved in cellular metabolism.

I thank the University of Western Australia for providing me with an environment conducive to the thinking, reading and discussing that were essential for the development of the book.

Finally I express my appreciation for one of my peers, the late Dr Ian Hansen, who was always willing, available and able to discuss the many ideas that arose during the writing of the lectures and the book.

Contents

List of Figures	ix
List of Boxes	xi
Introduction	1
1 Energy and the Cell	3
1.1 How does a chemical reaction provide energy?	3
1.2 How to access chemical energy within the constraints of cellular structure and function	4
1.3 Use of the free energy and the equilibrium constant of a reaction to quantify the energy available from a reaction	7
1.4 Energy in cellular fuels: where and how much?	8
1.5 How much energy is there in a reduced bond?	10
2 Oxidative Metabolism: the Reactions, the Electron Carriers and the Energy Broker	12
2.1 Oxidative metabolism comprises oxidation and preparation reactions	13
2.2 Electrons are transferred by biological electron carriers	13
2.3 ATP: chemistry to suit a role	15
3 Oxidative Metabolism: an Overview of the Pathways that convert the Reduced Bonds of Fuels to ATP	19
3.1 Glycolysis	20
3.2 β-oxidation	21
3.3 The Krebs cycle	22
3.4 The electron transfer system	23
4 Oxidative Metabolism: linking the Hardware with the Software	25
4.1 The electron transfer system: the ATP producer	25
4.2 The Krebs cycle: oxidation and CO_2 production	27
4.3 β-oxidation: providing reduced bonds from lipid for the Krebs cycle	31
4.4 Glycolysis: multi-purpose processing	33

		4.5 Energy content: carbohydrate vs lipid	37

5 Synthesising Reduced Bonds in Animal Cells: Fatty Acid Synthesis 40
   5.1 Fatty acid synthesis: the basic strategy 41
   5.2 Fatty acid synthesis: recurring concepts and themes, and a final message 41

6 Synthesising Reduced Bonds in Plant Cells: the Calvin Cycle and Photo-reduction 45
   6.1 The reactions involved in the synthesis of glucose: the Calvin cycle 45
   6.2 The synthesis of the $NADPH + H^+$ required for the reduction in the Calvin cycle 47
   6.3 The synthesis of the ATP required by the Calvin cycle 49

7 Reiteration, Variation and Application 51
   7.1 Figure I.1 revisited 51
   7.2 Variations of the basic scheme 52
      7.2.1 Producing end-products other than CO_2 and H_2O 52
      7.2.2 Using electron acceptors other than oxygen 53
      7.2.3 Using non-organic compounds as the source of electrons 54
   7.3 Application of the concepts: weight control in humans 55

Figures 59
Index of Concepts 115
Index of Repeating Themes 116
General Index 117

List of Figures

Figure I.1	The oxidation/reduction cycle in cells	61
Figure 1.1	The types of fuel molecules and their characteristics in terms of reduced bonds	62
Figure 1.2	The oxidation of glucose in terms of oxidation of reduced bonds	63
Figure 1.3	The oxidation of a reduced bond	64
Figure 2.1	Energy extraction and incorporation viewed as a reversible oxidation/reduction reaction	65
Figure 2.2	Examples of the three types of oxidation that occur in pathways of energy extraction	67
Figure 2.3	NAD and FAD, electron carriers	69
Figure 2.4	Standard reduction potentials of some biologically important redox pairs (V, 25 °C, pH 7.0)	70
Figure 2.5	The role of redox potential in electron transfer reactions	71
Figure 2.6	The structure of ATP and the thermodynamics of its hydrolysis	72
Figure 2.7	The role of ATP hydrolysis in enabling otherwise biologically impossible reactions	73
Figure 3.1	An overview of the pathways involved in the oxidation of carbohydrate and lipid fuels	75
Figure 3.2	A summary of glycolysis	77
Figure 3.3	A summary of β-oxidation	79
Figure 3.4	A summary of the Krebs cycle	80
Figure 3.5	A summary of the electron transfer system	81
Figure 4.1	A more detailed electron transfer system	83
Figure 4.2	ATP synthesis by the electron transfer system	85
Figure 4.3	Partitioning the Krebs cycle into functional units	87
Figure 4.4	The detailed reactions of the Krebs cycle	89
Figure 4.5	The structure of acetyl CoA	91
Figure 4.6	The detailed reactions of β-oxidation	93
Figure 4.7	The essentials of glycolysis	95

Figure 4.8	The three aldehyde oxidation reactions	97
Figure 4.9	Glycolysis in detail	100
Figure 4.10	The energy yield of the carbohydrate and lipid fuels	100
Figure 5.1	A simplified scheme of fatty acid synthesis: the synthesis of a 6-carbon saturated fatty acid	103
Figure 5.2	Fatty acid synthesis: recurring concepts and themes	105
Figure 6.1	The synthesis of reduced bonds in plants: the Calvin cycle	105
Figure 6.2	The chloroplast and the location of the elements of photo-reduction	109
Figure 6.3	Light energy, redox potentials and the proton gradient in chloroplasts	111

List of Boxes

Box 1.1	The equilibrium constant	4
Box 1.2	Calculating free energies of compounds and solutions	5
Box 1.3	Standard and actual free energy	8
Box 1.4	The equilibrium constant for the oxidation of glucose	9
Box 2.1	Repeating themes	12
Box 2.2	Calculation of ΔG from standard redox potentials	14
Box 3.1	Metabolic pathways are not set in stone	20
Box 3.2	Innocuous end-products	22

Introduction

The process of energy extraction and incorporation in cells comprises reactions which inter-convert cellular fuels and their breakdown or oxidation products. On an oxidised/reduced scale, the fuels (e.g. glucose) are reduced[1] and the breakdown products (CO_2 and H_2O) are oxidised. When fuels are oxidised, the energy made available[2] is utilised to do work in the cell (Figure I.1). Cells are continuously doing work to stay alive. They must, for example, replace proteins that are degraded, repair damaged DNA, maintain the concentrations of various ions against concentration gradients, and regulate the above and other processes.

The oxidised breakdown products are recycled through reduction to produce the fuels, but in this case energy input is required. Ultimately, all of this energy for reduction is solar energy, which is converted into a useable form by the process of photosynthesis in plants. Energy flow in and between cells is therefore a continuous cycle of oxidation and reduction; energy is continuously being harnessed to do work, and there is a constant input of solar energy (Figure I.1).

This book presents an orderly and increasingly detailed explanation of the various facets of Figure I.1, based on recurring concepts and themes. Chapters 1 and 2 provide the basic knowledge that is necessary to appreciate the design of the pathways discussed in the later chapters. Chapter 1 explains the nature of chemical energy, how it can be derived from the fuels via chemical reactions, and how it can be quantified. It also considers the nature of the fuels used in living cells, where the chemical energy in these molecules is located, and how much energy can be released. Chapter 2 describes the characteristics of biological oxidation and reduction reactions and the molecules that transfer electrons during these reactions. It also introduces the compound adenosine triphosphate (ATP; see Figure I.1), which is used to transfer small parcels of energy between the fuels and the reactions that use this energy. Chapter 3 is a general overview of the four pathways that comprise the process by which fuels are oxidised. This chapter allots each pathway a place and function. Chapter 4 provides greater detail of each of the pathways in Chapter 3, with constant reference to the recurring concepts and themes introduced in earlier chapters. Chapter 5 addresses fuel synthesis in animal cells, and demonstrates how the principles of oxidative metabolism apply equally to this reductive process. Chapter 6 is a description of the reduction specialist, photosynthesis. It again stresses how the principles of oxidation apply to reduction, but also demonstrates the limitation of reductive processes in animal cells. In the final chapter (Chapter 7) the major messages of the book are revisited with the benefit of hindsight. Also in this chapter, novel variations on the basic metabolic design are explored to show that hardware, but not software, can change. The final section of Chapter 7 is an application of the lessons learned to the major, yet simple problem of weight control in humans.

Notes

[1] For cellular fuels, the amount of energy available to the cell is determined by the degree of reduction of the fuel, which is related to the sum of the carbon-carbon and carbon-hydrogen bonds in the molecule. This is a concept and will be addressed in more detail later. Concepts form the backbone of this book. They will be formally identified and cross-referenced between chapters as the book proceeds (see Index of Concepts, p.115).

[2] Only a small fraction ($1/10^{10th}$) of the energy in the fuel molecules is available to the cell.

1 Energy and the Cell

1.1 How does a chemical reaction provide energy?

The concepts involved in thermodynamics can be difficult to grasp, and even more difficult to apply in a meaningful sense to a living cell. However, it is impossible to understand the principles of energy extraction, utilisation and incorporation in the cell without a basic understanding of this area. What follows is a list of points comprising the essential details that are necessary to understand the concepts that form the basis of this book. Where relevant, a constant temperature and pressure are assumed.

1. The amount of energy that is available to a cell in a molecule is related to how much energy is required to synthesise that molecule from its component atoms. This is termed the free energy of formation, $\Delta_f G°$. The $\Delta_f G°$ of a molecule is determined only by the electron interactions and bears no relation to the energy contained in the nuclei.

2. The total free energy of a solution of molecules is related to a combination of the sum of the $\Delta_f G°$ of the molecules in question,[1] and to the natural logarithm of the concentration of the molecule. This is a crucial point when considering reactions, as (a) the result of the natural logarithm component is that the total free energy of the reactants plus products can change with concentration, even if the total number of molecules stays constant (see point 4 below), and (b) the equilibrium constant[2] of the reaction is expressed in terms of the total free energy of the reactants and the products.

3. All reactions proceed towards the equilibrium position, at which point the total free energy of the reactants equals that of the products. At equilibrium the reaction between reactants and products is still occurring, but there is no change in the total free energy of the system, so the reaction is not able to do work. This is an important concept, reactions at equilibrium can do no work (concept: reactions at equilibrium and work).

4. Due to the effect of the logarithmic component of concentration on the total free energy of the reactants and products, when a reaction is out of equilibrium the total free energy of the system is greater than when the reaction is at equilibrium. This characteristic of reactions is the essence of energy availability in a cell, and introduces the next concept. As a reaction comes to

equilibrium, energy is lost from the system and can be used to do work (Figure I.1; concept: energy availability from a reaction out of equilibrium).

1.2 How to access chemical energy within the constraints of cellular structure and function

Again this issue is best addressed as sequential points.

1. For a reaction to produce energy so that work can be done, that reaction must be out of equilibrium (see 1.1, point 4).

2. The equilibrium position of a reaction is described by a mathematical function, the thermodynamic equilibrium constant (K_{eq}). This is the ratio of the concentrations of the reactants and the products of a reaction when that reaction is at equilibrium (Box 1.1). Remember that at equilibrium the total free energy of the reactants and products are equal (1.1 point 3).

Box 1.1   The equilibrium constant

K_{eq} = concentration of product(s)/concentration of reactant(s)

when the reaction is at equilibrium.

- When the reaction is at equilibrium, the total free energy of the reactants = the total free energy of the products.

---

3. For a reaction to be out of equilibrium there must be different total free energies on the two sides of the reaction. This can be accomplished through differences in the $\Delta_f G°$s of the reactant and product molecules, and/or by differences in the concentrations of the reactants and products.

Considering 1–3 (above), in evolutionary terms, what are the strategies that a cell can adopt in terms of designing reactions that are out of equilibrium and thus able to do work? Let us look at the options in 1.2 point 3 (above) in terms of a reaction between two molecules, A (reactant) and B (product), whose $\Delta_f G°$s are identical (an isomerisation reaction between glucose and galactose will suffice as an example even though the $\Delta_f G°$s may not be absolutely identical). Since these two molecules have the same $\Delta_f G°$, if they are present at the same concentration, the total free energy of the reactants and products will be the same and the reaction will be at equilibrium. The K_{eq} under these conditions will be 1.0 (Box 1.1 and 1.2, Scenario 1).

Box 1.2 Calculating free energies of compounds and solutions

The free energies of formation of molecules at 1 M and at standard temperatures and pressures ($\Delta_f G°$) are known and are available in standard reference books. The following two equations can then be used to work out the total free energies of solutions of these compounds, and the K_{eq} of reactions involving different compounds.

- (1) The chemical potential of A in solution (CP_A) = $\Delta_f G°_A$ + $RT(\ln [A])$. R is the universal gas constant (8.314 J·mole^{-1}·deg Kelvin^{-1}), T = °K.
- (2) The total free energy (G_A) of a litre of the solution of A in (1) = number of moles (N) × CP_A.
- Remember that at equilibrium, $G_{reactants} = G_{products}$.

So consider the following scenarios for the reaction A B:

SCENARIO 1

- $\Delta_f G°_A = \Delta_f G°_B = 1250$ kJ/mole (similar to the $\Delta_f G°_{glucose}$).
- [A] = [B] = 0.1 M (a feasible concentration of a substrate in the cell).
- G_A will equal G_B, so the K_{eq} = 0.1/0.1 = 1.
- Using equations (1) and (2) above, we can calculate that [A] has to be 99-fold higher than [B] for G_A to be 100-fold higher than G_B (i.e. the reaction is 100-fold out of equilibrium).

SCENARIO 2

- $\Delta_f G°_A$ = 12500 kJ/mole; $\Delta_f G°_B$ = 1250 kJ/mole.
- If [B] = 0.1 M, then G_A will equal G_B when [A] = 0.01 M.
- K_{eq} therefore = 0.1/0.01 = 10.
- Using equations (1) and (2) above, one can calculate that [A] only has to be 9.9-fold higher than [B] for G_A to be 100-fold higher than G_B (i.e. the reaction is 100-fold out of equilibrium).

In order for this reaction to produce some energy so that work can be done, the total free energy of the reactants must be greater than that of the products. Using equations (1) and (2) in Box 1.2, one can calculate that if there is to be ten times more total free energy in the reactants compared to the products (for the purposes of this book, this is termed 10-fold out of equilibrium), the concentration of A must be 9.9-fold higher than that of B. For the reaction to be 100-fold out of equilibrium, the concentration of A would have to be 99-fold higher than that of B. Note that when the reaction is 100-fold out of equilibrium, the equilibrium constant has not changed, it is still 1.0. But as the reaction is not at equilibrium it will tend towards equilibrium, and as it does so will release energy that can be used to do work. In this scenario (Box 1.2, Scenario 1), in which the $\Delta_f G^o$s of the reactants and products are the same, a large difference (99-fold) between the concentrations of reactants and products is required to get the system 100-fold out of equilibrium. Such a situation dictates that there has to be either a high concentration of A, or a low concentration of B. But it will be argued later that large differences in the concentrations of reactants and products, and either very large or very small concentrations of reactants and products, are not incompatible with cell structure and function.

So with these cellular constraints in mind, let us consider a reaction in which the $\Delta_f G^o$ of the reactant is 10-fold higher than that of the product. In this case the K_{eq} (when the total free energy of reactants and products are equal) is not 1.0, it is 10 (Box 1.2, Scenario 2), and the reaction would be 100-fold out of equilibrium when the concentration of A was only 9.9-fold higher than the concentration of B. It is clear from this scenario that a strategy of different $\Delta_f G^o$s for the reactants and products allows work to be done in a cell despite similar concentrations of reactants and products. The greater the difference in $\Delta_f G^o$ between the reactants and products, the smaller the concentration difference needed to get the reaction out of equilibrium.

Hence a final concept for this section can be formulated (concept: implications of high K_{eq}). The higher the K_{eq} of a reaction, the greater is the amount of work that can be done by that reaction, with only small (or no) differences between the concentrations of substrates and products.

Combining the three concepts above, and thinking in the context of a cell, we can conclude this section with the following two points.

1 Reactions must be out of equilibrium in order for work to be done.
2 There is only a limited amount of space in a cell, yet the cell must contain many complex structures (such as mitochondria), and many solutes (thousands of soluble proteins, and at least 500 other chemicals) in order to

be viable.[3] Therefore reactions with a high K_{eq}, i.e. those that are out of equilibrium with similar (which enables them both to be low as well) concentrations of reactants and products, are the most useful for doing work within the physical constraints of a cell. This is the most obvious advantage of reactions with a high K_{eq}, but there are other arguments for maintaining metabolite concentrations in the cell within the range of 10^{-6} M–10^{-3} M, which will not be discussed here (see Fell, 1997[4]).

1.3 Use of the free energy and the equilibrium constant of a reaction to quantify the energy available from a reaction

The relation between the equilibrium constant for a reaction and the energy available to do work as the reaction proceeds can be quantified using the concept of the free energy of a reaction. Free energy has been discussed previously, as it relates to individual compounds in solution. This idea will now be taken a step further with a new concept (concept: free energy of reactions) which is a useful one for quantitatively describing biological reactions. It is useful for two reasons:

- First, the standard free energy change for a reaction ($\Delta G°$) indicates whether the equilibrium constant of that reaction is large or small. As we have already seen, this is an informative characteristic of a biological reaction.

- Second, the $\Delta G°$ is used to calculate the actual free energy change (ΔG). The latter takes into account the concentrations of reactants and products that are found in the cell. ΔG therefore incorporates the K_{eq} (from the $\Delta G°$), and the cellular ratio of the concentrations of products and reactants, and allows the evaluation of the potential of all types of reactions to proceed towards equilibrium, and thus to do work, under conditions that actually occur in the cell (Box 1.3). A reaction will only proceed if its ΔG is negative.

So referring back to the reactions we considered earlier in Box 1.2, Scenario 1, since the K_{eq} was 1.0, the $\Delta G°' = -RT \ln 1.0 = 0$ kJ/mole.[5] This tells us that the K_{eq} (1.0) is small in this case. But even with a small K_{eq}, if this reaction is 100-fold out of equilibrium (i.e. [B]/[A] = 1 M/99 M = 0.01), the $\Delta G = 0 + RT \ln 0.01 = -11.4$ kJ/mole. Thus under these conditions, the ΔG for this reaction is large and negative and therefore the reaction is out of equilibrium and can do work. In this case a large difference in concentration between the reactants and the products is required.

However, when the $\Delta_f G°$ of A is 10-fold higher than that of B (Box 1.2, Scenario 2), the K_{eq} is no longer 1.0, but 10, and the $\Delta G°' = -5.7$ kJ/mole. This tells us

(a) that the K_{eq} is large, and (b) that a large and negative ΔG is possible at low and similar amounts of reactants and products. In Scenario 2, when the concentration of A is only ten times the concentration of B, ΔG = -11.4 kJ/mole.

Box 1.3 Standard and actual free energy

The standard free energy change for a reaction ($\Delta G^{o\prime}$) is related to the K_{eq} of that reaction by the equation:

$$\Delta G^{o\prime} \text{ (J/mole)} = -RT \ln K_{eq}$$

- ln K_{eq} is the natural logarithm of K_{eq}.
- When the K_{eq} is large, the $\Delta G^{o\prime}$ is large and negative.
- When the K_{eq} is 1.0, the $\Delta G^{o\prime}$ is zero.
- When the K_{eq} is less than 1.0, the $\Delta G^{o\prime}$ is positive.

The actual change in free energy (ΔG) is derived from the concentrations of substrates and products, and the standard free energy by the equation:

$$\Delta G \text{ (J/mole)} = \Delta G^{o\prime} + RT \ln [\text{products}]/[\text{reactants}]$$

A reaction will only proceed if its ΔG is negative. Therefore work can only be done by a reaction if the ΔG is negative, and the more negative the ΔG, the more work that can be done.

So in summary, we now know what sort of reactions can generate energy in the cell, and we know how to put a number on how much energy they can generate. The issue of energy in fuel molecules will now be addressed with this information in hand.

1.4 Energy in cellular fuels: where and how much?

The molecules used as fuels for cells can be grouped into three classes, carbohydrate, lipid and protein. The available energy is located in the C-C and C-H bonds[6] (Figure 1.1). There is a lot of energy in the nucleus of every atom in these molecules, but the oxidation of a mole of glucose only yields $1/10^{10\text{th}}$ of the energy that is available in the molecules. If our cells could extract all the energy in a glucose molecule, three micrograms of glucose would be enough to last us a lifetime! This becomes very clear when we see the results of nuclear fission in nuclear power stations and nuclear weapons.

Nuclear fission, however, is destructive, difficult to control, and does not occur under conditions conducive to cell viability. Therefore it is not a feasible mechanism for releasing energy in the cell.

Evolution has instead taken advantage of the concepts that have been discussed in sections 1.1 and 1.2 to release energy from these fuels. The mechanism comprises chemical change and ionisation, is feasible within the constraints of cellular chemistry, and involves the addition or removal of electrons. The nucleus is not involved, except in that it determines the number of electrons available. The disadvantage of this mechanism is that it can release energy from only certain parts of the fuel molecules, the reduced bonds. The mechanism embodies the two conclusions of section 1.2, i.e. work can only be done as a reaction comes to equilibrium, and a reaction with a high K_{eq} will be out of equilibrium even though the concentrations of the reactants and products are similar and low.

The reactions involved in this mechanism are oxidation reactions, whereby the fuel is oxidised to CO_2 and H_2O. At this point the practical implications of the concepts we have encountered begin to manifest. These oxidation reactions have staggeringly large equilibrium constants, as demonstrated for the oxidation of glucose in Box 1.4. The reactants can be defined as particular parts of the fuel molecules, i.e. C-H or C-C bonds, the reduced bonds referred to above and in Figure 1.1. The products are bonds with oxygen, such as H-O in H_2O and C-O (the C-O bonds occur transiently within the fuel as it is being oxidised, but ultimately are all found in CO_2; see Figure 1.2 and Figure 1.3). The reactants in this case represent vastly higher total free energies than the products, thus resulting in a reaction with an extremely high K_{eq}, which is out of equilibrium under all situations in the cell (Box 1.4).

Box 1.4   The equilibrium constant for the oxidation of glucose

$$K_{eq} = [CO_2] \times [H_2O]/[glucose] \times [O_2] = 10^{500,000}$$

This number is unimaginably large. To give you some idea of what this number means, this reaction is at equilibrium only when one molecule each of glucose and oxygen are contained in a carbon dioxide-saturated sphere of water fifty times the size of the solar system.

Therefore, as far as the cell is concerned, available energy is in the reduced C-

C and C-H bonds of the fuel molecules. This is the concept of reduced bonds (concept: reduced bonds). In the context of cellular energy, these bonds are considered reduced because the electrons in the bond are not being shared with oxygen. Electrons are removed from these reduced bonds to eventually become part of a bond with oxygen, which has a much greater affinity for electrons than either carbon or hydrogen[7] (hence the lower total free energy of the products of these oxidation reactions).

The other bond found in fuels is the C-O bond,[8] but it is worth no energy in a biological sense as the electrons in this bond are already part of a bond with oxygen. The value of any fuel molecule to a cell is therefore easily calculated by simply counting the number of C-C and C-H bonds. The similarities and differences between the fuel classes in Figure 1.1 now become clear. Carbohydrates and amino acids (the building blocks of proteins) contain similar amounts of available energy, in terms of the number of reduced bonds per carbon atom or per weight. The lipids such as palmitate contain between 1.5 and 2.8 times as many reduced bonds as the carbohydrates and amino acids, depending upon whether the calculation is done per carbon or per gram.[9] The practical implications of these differences also become clear. One can now appreciate why organisms that fly long distances, like migrating hummingbirds and some insects, carry lipid as their fuel rather than carbohydrate or protein. And if you want to lose weight, switch from bread and peanut butter (lipid) to the same weight of bread and jam (carbohydrate).

1.5 How much energy is there in a reduced bond?

In terms of an average day in the life of a human at rest, the energy from approximately 1.4×10^{22} reduced bonds is required each minute. This is about the equivalent of the power used by a 100 W light bulb. Figure 1.1 indicates that this number of bonds is contained in 0.35 g glucose, or 0.13 g of palmitate. Of course with intense exercise the need for energy increases. For example, the energy expenditure of an athlete running a marathon increases by about 20-fold. Under these conditions, one reduced bond is enough to propel the runner 1.3×10^{-21} m of the 42.2 km distance. The runner then requires the energy from 3.2×10^{25} or 32 billion, billion, billion, billion reduced bonds to complete the marathon course![10]

Notes

[1] The solvent also contributes, and so does the interaction between the solute and the solvent, but for our purposes it can be assumed that these contributions are a constant.

[2] The equilibrium constant (K_{eq}) and its significance is explained in section 1.2.

[3] There is a detailed, informative and fascinating discussion of solvent capacity by Atkinson, D E, *Cellular Energy Metabolism and its Regulation*, New York, Academic Press, 1977.

[4] Fell, D, *Understanding the Control of Metabolism*, London, Portland Press, 1997, section 1.4.2.

[5] All calculations are at 25 °C unless otherwise specified.

[6] This view of available energy in molecules was originally proposed by Stokes, G B, 'Estimating the energy content of nutrients', *Trends in Biochemical Sciences*, 1988, 13, pp.422–424.

[7] Oxygen is the element with the second highest electronegativity; fluorine has the highest.

[8] There are C-N and N-H bonds in amino acids, but these make up a trivial proportion of the total number of bonds in these molecules, and are not harnessed by oxidative metabolism.

[9] These numbers demonstrate the point in this case, but are not absolutely accurate, as not all reduced bonds are equal, as will become clear in Chapter 4. In terms of actual available energy, lipid contains 1.2 (per carbon) and 1.8 (per gram) times as much energy as carbohydrates and amino acids (Figure 4.10).

[10] As in footnote 9, these numbers demonstrate the point, but would vary according to the mix of carbohydrate and lipid that is used to fuel resting metabolism.

2 Oxidative Metabolism: the Reactions, the Electron Carriers and the Energy Broker

Figure 2.1 simplifies Figure I.1 into a reversible oxidation/reduction reaction. This reaction comprises a set of themes (Box 2.1) that appear repeatedly throughout the molecular processes associated with energy transfer. The reduction aspect of this figure will not be addressed until Chapter 5.

Box 2.1   Repeating themes

From this point on, various themes will appear repeatedly. This is because there are optimally effective and optimally efficient ways of accomplishing the various metabolic goals, and through the process of evolution these have been repeatedly selected for the task. These repetitions will appear in the form of similar, or even identical compounds, enzymes, pathways, and even structures which include all of the above. They appear in different compartments of the cell, in different cells, organisms and even phylogenetic kingdoms, and will be noted as they occur. They are instructive and useful as they convert metabolism from hundreds of seemingly unrelated enzymes, substrates and co-substrates, to a manageable and meaningful set of repeating themes.

It is the reduced C-C and C-H bonds of the fuels that are oxidised (so oxygen and the fuels are reactants), and the atoms that make up the fuels (C, H and O) appear as the products (CO_2 and H_2O). The next two sections of this chapter address some aspects of these oxidation reactions in detail, which will leave outstanding one aspect of the oxidative component of Figure 2.1, adenosine triphosphate or ATP. ATP has been mentioned previously (Figure I.1), but we are now sufficiently informed to discuss its structure, its $\Delta_f G°$ and the free energy changes associated with its formation and hydrolysis. This will be done in section 2.3, when its role as the link between reactions that are proceeding towards equilibrium and those that accomplish work in the cell is addressed.

2: OXIDATIVE METABOLISM

2.1 Oxidative metabolism comprises oxidation and preparation reactions

The oxidation of fuels requires two types of reactions, oxidation reactions and preparation or rearrangement reactions. With the oxidation reactions comes the first repeating theme (oxidation reactions; see Index of Repeating Themes, p.116); the oxidation reactions are always one of three types; the conversion of a CHOH group to a ketone, an alkane to an alkene, or an aldehyde to an acid (Figure 2.2). Whenever one of these reactions occurs in the oxidative (energy extraction) pathways of energy metabolism, a C-H bond is oxidised and the total number of reduced bonds in the product is one less than that in the reactant. Note at this point that in terms of a cell, C-C bonds are worth as much as C-H bonds, but they are not directly oxidised by any of the reactions depicted in Figure 2.2. Therefore, and this is another repeating theme (preparation reactions), some of the reactions in the oxidative pathways are preparation reactions. One type of preparation reaction converts C-C bonds to C-H bonds by rearranging the molecule. The C-H bonds are then oxidised. We will return to preparation reactions when the details of the pathways are considered in Chapter 4.

The electrons removed from the C-H bonds are eventually shared with oxygen (Figure 1.3), but this does not happen directly. They are first transferred to dedicated electron-carrying organic compounds, that in turn eventually transfer the electrons to molecular oxygen.

2.2 Electrons are transferred by biological electron carriers

There are many electron carriers involved in producing ATP from reduced bonds. Only the ones directly involved in accepting the electrons from the C-H bond of the fuels will initially be considered. These are $FADH_2$ and $NADH + H^+$ (Figure 2.3). With the appearance of these two compounds comes the introduction of another concept and three more repeating themes.

First the concept. Electron carriers have an affinity for electrons that is determined by their chemical structure. Hence the concept (redox potential) of quantifying the tendency of the electron carriers to take up (which is related to the tendency to donate) electrons. An electron carrier has an oxidised and a reduced form. The affinity for electrons of the oxidised form can be expressed as an electrical potential, the oxidation-reduction (or redox) potential, with units of volts or millivolts. The theory behind these measurements can be found in any comprehensive biochemistry textbook. For the purposes of our discussion it is enough to know that the higher the redox potential the greater the affinity for electrons of the oxidised form of the redox pair.

Some oxidised/reduced pairs are shown in Figure 2.4, with their standard redox potentials (E'_0).[1] As with $\Delta G^{o\prime}$, standard redox potentials per se can provide meaningful information, e.g. NADH + H$^+$ should be able to donate electrons to oxygen, and this is exactly what happens in the electron transfer system that is described in Chapter 4. But as with non-redox reactions, the concentrations of reactants and products can play a role in determining whether a reactions will proceed in the cell. In this respect the standard redox potential can also be used in the calculation of ΔG for reactions with a redox component. For example, from Figure 2.4, one would deduce that NAD could not accept electrons from the CHOH group of lactate and donate them to oxygen, as is depicted in Figure 2.5. But this happens in the cell because the ΔG (which takes account of the concentrations of products and reactants) for this reaction is negative (Box 2.2 and refer back to Box 1.3).

Box 2.2 Calculation of ΔG from standard redox potentials

The ΔG for the reaction

$$\text{lactate} + \text{NAD} \rightarrow \text{pyruvate} + \text{NADH} + \text{H}^+$$

is calculated as follows, and would be negative whenever this reaction is occurring.

- The redox potential (E) for each component redox reaction
  $= E'_0 + RT/n\mathcal{F} \ln [\text{electron acceptor}]/[\text{electron donor}]$

  n = number of electrons transferred, $\mathcal{F}$ = Faraday's constant (J/volt × mole)

- ΔE for the combined reaction = E (lactate/pyruvate reaction) - E (NAD/NADH reaction).

- ΔG for the combined reaction = $-n\mathcal{F}\Delta E$.

The first repeating theme that comes with the introduction of FAD and NAD, is the 'business-end' theme. Note how large both these molecules are: the molecular weights are in the order of 700. These molecules comprise a number of different types of structures, compared for example with the relatively uniform structure of glucose, which has a molecular weight of 180. Yet the electron-carrying function is accomplished by a small part of these molecules, perhaps four atoms out of the total of about seventy that comprise the molecule. The rest (90%) of the molecule is multifunctional. Some parts, the nicotinamide and isoalloxazine rings, endow the molecule with a specific affinity for electrons. Other parts, such as the adenosine monophosphate (AMP), provide a molecule that binds specifically and with the right affinity to the enzymes involved in the oxidation reaction.[2]

The second repeating theme that arises with NAD and FAD is that of structural repeats (structural repeats: AMP) Note that both NAD and FAD contain AMP. You will come across this same structural theme in other compounds later. AMP in NAD and FAD plays no role in redox potential, nor does it function like the related molecule, ATP. It plays a central role in the binding of NAD and FAD to enzymes and proteins with which they interact.[3] As discussed in Box 2.1, these structural repeats bring meaning, order and predictability to a potentially confusing mix of unrelated proteins, compounds and reactions. The third repeating theme (NAD and FAD specificity) is one of NAD and FAD specificity for particular types of oxidation reactions. This theme is detailed in Figure 2.3.

We now understand reduced bonds, electron carriers and how these two interact in a reaction with a negative ΔG to oxidise the reduced bonds, ultimately donating the electrons to oxygen to produce energy. But so far no energy has been produced in a form that can be used by the cell to do the work defined in Figure I.1. There are many steps between the actual oxidation of reduced bonds and the production of useable energy. All of these are instructive as they reinforce the concepts and repeating themes that have been encountered up to this point. But before we embark on the stepwise journey through oxidative metabolism, the energy broker (ATP) that is actually produced at the end, will be introduced. This is the energy mentioned earlier that is accessible and useable by the cell.

2.3 ATP: chemistry to suit a role

Not only does a consideration of ATP show how a molecule can evolve to suit a role, it also brings together a number of themes that have been encountered previously. This is an important stage in the understanding of metabolism, as for the first time the reader has enough previous knowledge to understand the function of a previously unknown compound. This represents a watershed, as it demonstrates that metabolism is not (as is often the perception) an overwhelming list of new compounds and reactions that have to be rote learned. Rather, it is a series of variations on a few themes. If metabolism is seen in this light, it becomes easier to understand and learn, and it becomes more interesting. From now on you will increasingly encounter familiar reactions, structures and concepts.

The free energy released when electrons from the reduced bonds are donated ultimately to oxygen is not coupled directly to reactions and processes that require this energy. Instead the free energy is initially conserved as ATP.[4] The ATP hydrolysis reaction is then used in an ingenious way (as explained in more detail below), with the result that reactions that previously had ΔGs that were close to zero, or positive, become part of a spontaneous reaction with a large and negative ΔG (see Figure I.1). The ATP molecule will now be addressed in detail. With the benefits of the concepts and repeating themes that have already been introduced, we will develop an understanding of why this molecule is suited to its role.

- The structure of ATP demonstrates the theme of structural repeats (structural repeats: AMP; Figure 2.6). Like the structures of NAD and FAD, ATP contains AMP. And like NAD and FAD, the AMP part of the molecule plays a role in binding to enzymes. But the unique attribute of ATP is the equilibrium constant of its hydrolysis reaction.

- ATP can be hydrolysed to ADP and inorganic phosphate (Pi) by the hydrolysis of the terminal phosphate anhydride bond (Figure 2.6). The $\Delta_f G°$ of the ATP molecule is vastly higher than the sum of the $\Delta_f G°$s of its hydrolysis products. So as argued in section 1.2, the Keq of this reaction will be large, and the $\Delta G°'$ of the reaction is large and negative. Consequently, as also argued in section 1.2, even though ATP and its hydrolysis products are found at similar, and relatively low concentrations in the cell,[5] the ratio of products/reactants in the cell is approximately 10^9 times lower than it would be at equilibrium. Accordingly the ΔG of the reaction in the cell is large and negative (Figure 2.6). ATP hydrolysis can therefore be used in the cell for two reasons.

  1 To 'drive' reactions that would otherwise not proceed. How this is accomplished involves a new concept (using ATP hydrolysis as part of a combination reaction with a large and negative ΔG). ATP hydrolysis is incorporated as part of a new, combination reaction, therefore converting the positive ΔG of the original reaction to a negative one for the combined reaction (Figure 2.7).

  2 To create a combined reaction with a high K_{eq}, which will proceed in the cell according to metabolic needs, despite perturbations of the concentrations of its reactants and products. The original reaction may have a K_{eq} greater than 1.0 in the cell, but not high enough to be insensitive to the concentrations of its reactants and products.

These strategies have obviously required the evolution of enzymes that (a) will bind the primary substrate, (b) bind ATP as well, and (c) catalyse a reaction that involves both substrates.[6]

Take a moment to reflect on why ATP is suited to its role. Any reaction with a cellular ratio of products/reactants that was 10^9-fold lower than the ratio at equilibrium, with therefore a large negative ΔG, could serve the same purpose, i.e. 'driving' unfavourable reactions. For instance the isomerase reaction converting glucose to galactose (the reactant and product have similar $\Delta_f G^o$s, and the K_{eq} of the reaction is approximately 1.0), if out of equilibrium to the same extent could fulfil the same function as the ATP hydrolysis reaction. But assuming the concentration of galactose in the cell was 0.2 mM (similar to that of ADP), the glucose concentration would have to be 0.2×10^7 M for this reaction to be 10^9-fold out of equilibrium. This much glucose would not dissolve in water, let alone fit into a cell! On the other hand the strategy could be to leave the glucose concentration at 3 mM, in which case the galactose concentration would have to be 3×10^{-7} mM or 0.3 nM. In this case, as well as the problem of the enzyme actually finding a galactose molecule with which to interact, the enzyme would be tricky to design, with affinities for its two substrates that differed by 7 orders of magnitude. So ATP has special qualities. K_{eq} or $\Delta G^{o\prime}$ are the key properties of the ATP hydrolysis reaction, and these are determined by the chemistry of the substrates and products.

Notes

[1] Each solute at 1 M, pH 7.0.

[2] This principle is at its most obvious in the case of protein enzymes, where in some cases only three amino acids out of 600 are involved in the catalytic function of the enzyme. The function of many, sometimes all of the remaining amino acids, is to get the catalytic amino acids into exactly the right configuration in space.

[3] The repeated presence of AMP may be related to the fact that early enzymes were made of RNA, and compounds with AMP in their structure would have been able to bind to these enzymes by base pairing with uracil.

[4] It is important to realise that this can be a misleading expression, as explained by Atkinson, (1977; Chapter 1, note 3). Free energy is not a physical entity in the same sense as heat or mass. Free energy is a function of concentrations and the free energy change of a reaction is not fixed, but depends upon concentrations. So to be strictly correct, and in keeping with concepts previously encountered, one would have to say that part of the energy from bond oxidation is conserved in the maintenance of a non-equilibrium ratio of ATP and its hydrolysis products.

[5] Concentrations will vary with cell type and situation, but these are some representative values. [ATP] = 2.5 mM, [ADP] = 0.2 mM, [Pi] = 1.6 mM.

[6] It is important to remember that 'driving a reaction' is an oxymoron, and a potentially misleading way to describe the use of the free energy of ATP hydrolysis; no reaction can proceed unless the ΔG is negative. But 'driving' is a useful way of describing the net result of incorporating ATP hydrolysis into a combination reaction.

3 Oxidative Metabolism: an Overview of the Pathways that convert the Reduced Bonds of Fuels to ATP

Figure 3.1 is an overview of the pathways involved in the complete oxidation of carbohydrate and lipid fuels. A general diagram is a useful beginning as it highlights patterns in the scheme, and reiterates some of the concepts and themes that were encountered in earlier chapters. These are listed below.

- There are two sources of reduced bonds: carbohydrate and lipid. If protein is used as a fuel, it is essentially converted into carbohydrate before it is oxidised.

- Note that the overall equation for carbohydrate oxidation is identical to that shown in the oxidative component of Figure 2.1. Carbon, hydrogen and oxygen are the reactants and CO_2, H_2O and ATP are the products. The only difference between a wood fire (see Figure 2.1) and the process depicted in Figure 3.1, is that the latter comprises at least thirty reactions between the reactants and the products.

- Both carbohydrate and lipid metabolism are characterised by a unique preparation phase (glycolysis and β-oxidation respectively). Both produce the same product, acetyl CoA (acetyl coenzyme A). CoA will be examined in detail in Chapter 4; for now it is enough to know that it carries two carbons and the associated reduced bonds from lipid and carbohydrate fuels (see next point). The oxidation pathways of carbohydrates and lipids are common after the production of acetyl CoA.

- The acetyl group of acetyl CoA comprises two carbons and four reduced bonds, and represents the majority of the reduced bonds of the lipid and carbohydrate fuels. It is oxidised, and degraded to CO_2 in the Krebs cycle (see section 3.3). This is where the carbon and the oxygen of the fuels exit the oxidative process.

- Some reduced electron carriers ($FADH_2$ and $NADH + H^+$) are produced by the two preparation phase pathways (glycolysis and β-oxidation), but most are produced in the Krebs cycle.

- The reduced electron carriers are oxidised in the electron transfer system. This is where most of the cell's ATP is produced under aerobic conditions, it is where the final substrate in the oxidation reaction (oxygen) is used, and

is where the hydrogen of the fuels exits the oxidative process as water.
- ATP can be produced by glycolysis without the involvement of oxygen. This is a crucial capability of a cell. It may seem contrary to the importance that has been placed on oxygen as an electron acceptor, but as will be explained later, the principles involved remain the same, although the efficiency is low.

The four pathways shown in Figure 3.1 will now be briefly addressed in turn. Each pathway will be linked to a function. It will become clear how much energy is extracted, and in what form this energy is extracted, at each point in the scheme shown in Figure 3.1.

3.1 Glycolysis

The functions of glycolysis are twofold.

- To degrade glucose to the acetyl moiety of acetyl CoA so that complete oxidation can take place by the next stage of energy metabolism, the Krebs cycle.
- To produce ATP in the absence of oxygen (Figure 3.2). One mole of a 6-carbon sugar, glucose, is processed by a sequence of about ten different reactions, to two moles of a 3-carbon compound, pyruvate. Pyruvate has three possible fates in most cells, but there are other options in the cells of some organisms (Box 3.1 and Figure 3.2). Oxygen is not involved in any aspect of glycolysis.

Box 3.1 Metabolic pathways are not set in stone

Organisms inhabit an impressive range of habitats. From the very hot to the very cold, from the dry to the wet, from the low to the high pressure, and from the aerobic to the anoxic. There is an equally impressive array of variations on the 'standard' metabolic themes, as energy extraction mechanisms have evolved to cope with, or to take advantage of, particular environments. For example, some cells never use oxygen as the final electron acceptor, as it is not available. Likewise some cells produce ATP in the absence of oxygen by glycolysis, but combine the pyruvate with arginine (because it is available from another process occurring simultaneously) to produce an end-product that suits the circumstances.

If glycolysis is considered in the context of its role as the initial preparation of glucose for oxidation, pyruvate crosses the mitochondrial membrane and is then converted to acetyl CoA. In the process, approximately a third of the energy available in glucose is extracted as a combination of actual ATP (about 6%) and reduced electron carriers (about 27%). Further metabolism of the acetyl CoA results in the oxidative extraction of the remainder of the available energy from the initial glucose molecule.

Glycolysis will now be considered in the context of its role as an ATP-producing pathway in the absence of oxygen. In this scenario the pyruvate is converted in the cytosol to either ethanol (which is produced, for example, by yeast in the brewing of beer) or lactate (which is produced in your muscles during a 100 m sprint). Approximately 6% (as mentioned above) of the energy available in glucose is extracted as ATP as a result of this process and comprises the *only* ATP that can be produced (albeit inefficiently) in the absence of oxygen. This function of glycolysis will be revisited in more detail in the next chapter. The next logical step in this general overview is to bring lipid oxidation to a common point with carbohydrate oxidation (refer back to Figure 3.1), in the form of acetyl CoA. This is done by β-oxidation.

3.2 β-oxidation

β-oxidation consists of four reactions in the mitochondria that repeatedly oxidise the β-carbon on a progressively shorter fatty acid molecule. β-oxidation produces the same 2-C unit (the acetyl moiety of acetyl CoA) as is produced when glycolysis is functioning as the preparation pathway for the complete oxidation of carbohydrate (Figure 3.3). The fuel in this case is a long-chain fatty acid which is processed two carbons at a time. The production of the acetyl CoA unit involves two oxidations (which conform to the rules outlined in Figure 2.2) and the conservation, as reduced electron carriers, of about 15% of the energy available in the two carbon portion of the lipid fuel. As in glycolysis, oxygen is not involved in β-oxidation. But in contrast to glycolysis, no ATP per se is produced by β-oxidation. This has major implications for the cell, as lipids cannot be used as a source of energy in the absence of oxygen.

It is timely at this stage to reiterate that up to the stage of acetyl CoA production, the only ATP that has been produced is that in glycolysis. 15–27% of the available energy has been extracted as reduced electron carriers, but under aerobic conditions, glycolysis and β-oxidation should be considered as pathways whose major function is to prepare the fuel molecule for complete oxidation in the Krebs cycle.

3.3 The Krebs cycle

The Krebs cycle (also called the tricarboxylic acid or citric acid cycle) occurs in the mitochondrial matrix (Figure 3.4). Its substrate is the acetyl CoA produced by glycolysis and β-oxidation. The acetyl (2-carbon) unit of acetyl CoA is condensed with a 4-carbon compound and the seven subsequent reactions accomplish the three functions of the cycle, which are:

- to oxidise the reduced bonds of the 2-C unit and conserve the energy as reduced electron carriers.
- to produce an innocuous and excretable end-product (CO_2) from the carbon and oxygen of the original fuels (concept: innocuous end-products; Box 3.2).
- to produce another 4-C compound to combine with the next incoming 2-C unit.

After the Krebs cycle, all that is left of the original fuels are the electrons from the reduced bonds. These are stored temporarily in the electron carriers $NADH + H^+$ and $FADH_2$, which account for the majority (about 90%[1]) of the available energy originally present in the fuels. All that remains to be done is to convert the energy, now conserved as reduced electron carriers, into ATP. This is carried out by an electron transfer system.

Box 3.2 Innocuous end-products

Oxidative metabolism for most cells is a constant process with a high rate of turnover. For example, your brain accounts for about 20% of the oxygen used by your body, and this rate of utilisation is constant and necessary for life. It is therefore essential that the end-products of this process in no way limit the process itself. CO_2 satisfies this criterion perfectly. CO_2 can be lost across lungs or gills or other epithelial surfaces to the vast atmospheric sink which comprises only about 0.03% CO_2. CO_2 therefore does not constitute a potential pollution or accumulation problem. Water is the other end-product of oxidative metabolism, ultimately produced by the electron transfer system (section 3.4). It also satisfies the end-product criteria as most organisms comprise 60–80% water, and the concentration of water in the cell is approximately 50 M. So water is certainly not a polluting end-product, and does not constitute an accumulation problem.

3.4 The electron transfer system

The electron transfer system (ETS) comprises a series of molecules in the inner mitochondrial membrane, which are electron carriers. As discussed in section 2.2, these molecules have the ability to accept and to donate electrons, and they have a specific affinity for electrons. The electrons from the reduced electron carriers (originally from carbohydrate and lipid fuels via glycolysis and β-oxidation) are transferred through the series of electron acceptors/donators which are organised such that each carrier has a higher redox potential than the carrier before it in the sequence (see Figure 2.4). The electrons are finally donated to oxygen, which has the highest redox potential, to form H_2O, with the concomitant formation of ATP.

- The process consists of seven major steps and produces about 90% of the ATP that is ultimately formed from the fuels (Figure 3.5). How ATP is made during this process entails a mechanism which is unimaginable to you at this stage. It is however based on concepts outlined in Chapter 1, and will be explained in Chapter 4. Take note that this mechanism is considered by an eminent biochemist to be 'the most important energy transduction mechanism in the biosphere' (Lehninger, 1993).[2]

- The substrates of the electron transport system are the reduced electron carriers produced by glycolysis and β-oxidation, ADP, Pi and oxygen. Its products are the oxidised electron carriers, ATP and H_2O.

- The last element of the fuels (hydrogen) makes its appearance in H_2O (see Box 3.2), and the role of oxygen in energy production is explained, along with the fact that you spend all your time breathing.

With the help of these summaries of the individual pathways, look back at Figure 3.1. It should now make more sense as a detailed version of the equation for the oxidation of glucose.

You are now somewhat of an expert on oxidative metabolism. You understand the oxidative software, i.e. the concepts and repeating themes that are the basis of the mechanisms of oxidative energy extraction and utilisation. You are familiar with, at a general level, the oxidative hardware, i.e. the four pathways that extract energy from the reduced bonds of the fuels. From now on there is very little that is new, but this repetition is not tedious, it illustrates the basic principles coming to fruition as functioning oxidative pathways.

In Chapter 4, the hardware will be explored in greater detail, to demonstrate at the level of individual reactions, the concepts and themes that have already been encountered. There will be a few new ideas, but these will mostly be

3: OXIDATIVE METABOLISM

sub-concepts, i.e. concepts within the framework of concepts we have already mentioned. Similarly with Chapters 5 and 6, there will be new pathways and new reactions, but mostly familiar concepts or themes. So sit back and enjoy the rest of the book, as all those previously confusing reactions and structures fall neatly into (as the title of the book suggests) one of a finite number of recurring concepts and/or themes.

Notes

[1] You may find this number confusing, as the only other energy output that has been identified is the 6% as ATP production per se in glycolysis. But there is another actual ATP produced in the Krebs cycle, so the percentage stored in the reduced electron carriers is somewhere between 91 and 94. The ATP produced in the Krebs cycle has not been mentioned in the text up to this point, as it has no significant bearing on the quantitative aspect of energy production. The reaction involved, however, is the essence of a crucial concept that we will encounter in Chapter 4.

[2] Lehninger, A L, *Principles of Biochemistry*, New York, Worth Publishers, 1993.

4 Oxidative Metabolism: linking the Hardware with the Software

In this chapter, the pathways will be covered in reverse order. This is done so that the chapter begins with the whole point of the system, i.e. the synthesis of ATP, and works back from the electron transfer system to the reduced bonds that are the original source (in an animal cell) of energy.

4.1 The electron transfer system: the ATP producer

Electron transfer systems occur on and in membranes. In prokaryotic organisms (bacteria) these are the plasmalemma or endoplasmic membranes. In eukaryotes the membranes are the inner membranes of mitochondria and chloroplasts. The intricate details of the components and reactions of the electron transfer system in animal cells are available in biochemistry textbooks, we have no need to go into this sort of detail. An adequate (for our purposes) scheme of the electron transfer system is shown in Figure 4.1. This figure is a more detailed version of Figure 3.5 and is adequately explained in the legends of Figures 3.5 and 4.1, and in section 3.4. But keep in mind where we are in the scheme; look back at Figure 3.1. The electrons involved come from the reduced electron carriers, and originally from the C-C and C-H bonds of the fuel molecules.

The first issue to be addressed is the synthesis of ATP, the molecule that was first mentioned in Figure I.1, the molecule that was discussed in detail in section 2.3, and the molecule whose synthesis is the *raison d'être* of the overall scheme outlined in Chapter 3. This is the mechanism that was deemed unimaginable in section 3.4, and so it is. So you have not encountered such a mechanism before, but you have encountered all the concepts and themes before. Each electron transfer in Figure 4.1 can be considered as a reaction coming towards equilibrium, as there is more total energy in the electron donor than in the electron acceptor. So energy becomes available as these reactions proceed (revise the concept: energy availability from a reaction out of equilibrium; Chapter 1). The transfers are arranged so that in addition to transferring electrons, the energy that becomes available is used to drive a net transport of H^+ across the membrane (Figure 4.2). In this way the relatively impermeable inner mitochondrial membrane becomes polarised, with a higher concentration of H^+ (lower pH) on one side than the other. In thermodynamic terms the change in free energy as the electron transfer reactions proceed towards equilibrium is conserved as this gradient across the membrane.[1] The protons flow back into the mitochondria (another reaction coming to equilibrium) through a proton channel, which comprises an enzyme which is designed to use the energy released to reverse the reaction described in Figure 2.7 (reaction 3), i.e. it synthesises ATP and is called an ATP synthase. This enzyme has a molecular mass of 515 kDa, and comprises eight types of subunits and about twenty actual subunits. An appropriately impressive final component of a remarkable process – and surely, one would think, unique in structure and function (see below)?

The identities of the substrates and products of the electron transfer system, and the processes and components that comprise the system have now been covered. ATP has been synthesised using energy originally residing in the reduced bonds of the carbohydrate and lipid fuels. The remainder of this section will be used (a) to show how this system demonstrates the concepts and themes that have been previously encountered, and (b) to identify any new or sub-concepts or themes that arise as part of this particular process.

The electron transfer system (ETS) demonstrates all of the concepts that relate to free energy, equilibria and reactions, and is effectively the embodiment of the concept of the energy available to the cell in reduced bonds. But the ETS is also the source of some instructive reiterations, and new sub-concepts and themes that are discussed in more detail below.

- The ETS is the classic example of the role of redox potentials (see section 2.2). Due to the nature of the reactions and compounds involved, concentrations are no longer the major issue. In most of the reactions in the ETS, the redox potentials of the donors and acceptors are the main determinants of the ΔG of the reaction. This is evident in Figure 4.1, in which the redox potentials show a consistent increase from NAD/NADH + H^+ to complex IV. Oxygen, the final acceptor, has the highest affinity for electrons (see Chapter 1, note 7).

- A sub-concept that arises in the ETS is related to that of linking favourable reactions to unfavourable reactions to produce a combination reaction that is favourable (see concept: using ATP hydrolysis as part of a combination reaction with a large and negative ΔG). The flow of protons down a gradient becomes part of a reaction that accomplishes the reversal of the reaction that hydrolyses ATP. The latter has a positive ΔG in the cell under all reasonable conditions (Figure 2.6), but in this case becomes part of a combination reaction with a negative ΔG.

- The components of the ETS are good examples of the repeating theme: business-ends. Refer back to Figure 4.1. The components of the system are mostly large proteins, yet the electron accepting and donating is done by small compounds (FMN) or even individual atoms (Cu). The role of the protein is to position the complex in the membrane, and to provide an environment that produces the appropriate redox potential of the electron acceptor.

- The point above introduces a new repeating theme, that of the co-opting of non-protein specialist groups by proteins (repeating theme: prosthetic groups). Note that in none of the components of the ETS do amino acids take on the task of accepting and donating electrons. Proteins are flexible and adaptable, but there are tasks that even proteins, with their almost infinite variability, cannot perform as effectively as non-protein molecules. Therefore specialist molecules/atoms (prosthetic groups) are co-opted for the task, a pattern seen repeatedly in the protein components of the various pathways that will be addressed in subsequent chapters.

- And finally, to address a question raised earlier concerning the ATP synthase being unique: the answer, amazingly, is no! It is simply another example of a structural repeat; we will come across a very similar (in structure and function) protein in another kingdom (the plants) when we consider photosynthesis in Chapter 6.

4.2 The Krebs cycle: oxidation and CO_2 production

Before we begin, get your bearings, look back again at Figure 3.1. NADH + H^+ and $FADH_2$ have been oxidised by the ETS and we will now look in detail at the production of these reduced electron carriers by the Krebs cycle. The Krebs cycle is where the majority of the reduced bonds from the fuels are oxidised, and it is the clearest, simplest, and most satisfying demonstration of concepts and themes that will be encountered. So it will be treated in detail, but rest assured that you have seen most of it before.

First, look at Figure 4.3. This figure splits the cycle into segments and emphasises how it works. Reduced bonds enter the cycle, and the subsequent pattern is a series of rearrangements (see repeating theme: preparation reactions; Chapter 2), oxidations and decarboxylations. The end result is the oxidation of four reduced bonds, and the production of four reduced electron carriers, two molecules of CO_2, and the original 4-carbon acceptor of the acetyl group.

Now move on to Figure 4.4, which outlines the reactions and structures in detail.

- The entry compound is acetyl CoA. You encountered this compound in Chapter 3, it is the point at which the oxidation pathways for carbohydrate and lipid converge. The acetyl group is carrying 70–85% of the reduced bonds of the original fuels and we will return to it shortly. But CoA is a goldmine of reiterated concepts and repeating themes, so will be examined in some detail before proceeding.

    1 CoA reiterates the concepts associated with ATP. Remember that the $\Delta G^{o\prime}$ of the ATP hydrolysis reaction, involving the terminal phosphate anhydride bond of ATP, is large and negative. The same is true for the thioester bond that links the acetyl group to CoA (see Figure 4.5). Acetyl CoA is used as a reactant in a combination reaction (section 2.3; concept: using ATP hydrolysis as part of a combination reaction with a large and negative ΔG). The result of this is that the entry reaction to the Krebs cycle, which involves the hydrolysis of acetyl CoA, will proceed despite very low levels of its substrates (acetyl CoA and oxaloacetate). The rationale for the low concentrations of these substrates is related to complex issues of metabolic regulation, which are not addressed in this book. But the structure of acetyl CoA is a good example, and a reminder, of how a reaction with a high K_{eq}, and a large and negative $\Delta G^{o\prime}$ can be used to ensure that a reaction will proceed under most conditions in the cell.

    2 CoA is also an example of a repeating theme (business-ends). It is a large and complex molecule, yet it is only the SH group that is involved in the binding of the acetyl group.

    3 The rest of the molecule serves two purposes. The entire phosphopantetheine group is an integral part of the acyl binding capacity of the molecule. This function is highlighted by the fact that you will encounter this group again in a synthesis pathway that is considered in Chapter 5. So the phosphopantetheine group is an example of a structural repeat (phosphopantetheine). The AMP moiety of CoA serves an enzyme binding function, so it too represents another structural repeat (AMP) and was discussed initially in section 2.2 and in Chapter 2, note 3). So do not look at acetyl CoA as a new compound; consider it as another manifestation of the familiar concepts and themes that can make metabolism easy to grasp and understand.

- We will now move on to reaction 1 in the Krebs cycle. It has a high K_{eq} (as discussed above), and it incorporates into citrate, the two carbons and the four reduced bonds of the acetyl group of acetyl CoA. Make sure you understand what has happened. The methyl group of the acetyl group has condensed with the ketone group of oxaloacetate. This reaction introduces a new repeating theme, that of recycling. As mentioned previously, there are some compounds that are in very low concentrations in the cell. These compounds are therefore recycled. CoA is one of these, and it can be seen in this first reaction that the CoA is released, and will be recycled, i.e. it will pick up another acetyl group from the fuels and deliver it to the Krebs cycle. You will encounter a more pressing example of the need for recycling in glycolysis later in this chapter.

- We now have the citrate molecule, which represents a quandary for the Krebs cycle, the purpose of which is to oxidise reduced bonds. Remember that we have learned that the oxidation of reduced bonds takes place by only the three reactions that are outlined in Figure 2.2, in which the substrate is a CHOH group, an aldehyde or an alkane. But the structure of citrate does not contain any of these groups, so by definition cannot be oxidised. The cycle therefore by necessity at this point involves a preparation reaction (see repeating theme: preparation reaction). Reaction 2 in Figure 4.4 (actually a sequence of two reactions) rearranges citrate into isocitrate, which does contain a reduced bond that can be oxidised; isocitrate contains a CHOH group.

- Reaction 3 is therefore the first oxidation – of an alcohol (isocitrate) to a ketone (α-ketoglutarate) – as depicted in Figure 2.2. An alcohol oxidation always involves NAD, so NADH + H^+ is produced (this is the repeating theme: NAD and FAD specificity). Note that the reduced bond involved is not one of those in the acetyl group that just entered the cycle.[2] But reaction 3 is also a decarboxylation, as this reaction is an oxidative decarboxylation. In this reaction one of the two carbons that are lost in each cycle (to account for the two carbons that enter as the acetyl group) is lost as CO_2, which is one of the innocuous end-products discussed in Chapter 3 and Box 3.2 (concept: innocuous end-products). Note that the CO_2 released contains no reduced bonds, so is therefore of no further use to the oxidative process. But the decarboxylation aspect of this reaction reiterates another element of the preparation reaction theme that was discussed in section 2.1 (repeating theme: preparation reaction). Note the C-C bond between the carbon that will be lost during decarboxylation, and carbon 3 of isocitrate. This is a valuable reduced bond, but cannot be oxidised, so is converted into a C-H bond as part of the decarboxylation reaction. Take special note of this, as we will identify the oxidation of this very bond later in the cycle. So reaction 3 is an informative one, being a combination of an oxidation, a decarboxylation and a preparation reaction.

- Reaction 4 is one of the most conceptually difficult reactions you will encounter, and is explained in points 1 and 2 below. It is catalysed by a very large protein complex (MW = 6 million kDa), comprising three different enzymes and three prosthetic groups. The reaction introduces a new concept, but this concept is repeated twice in glycolysis in very similar reactions, so at least you get a triple exposure to this one.

    1   Note that the substrate of this reaction (α-ketoglutarate) does not contain an oxidisable carbon. So a preparation step is included in this reaction (repeating theme: preparation reactions). A decarboxylation produces CO_2 from the carboxyl group of carbon # 1, and as in the previous reaction the C-C bond is conserved as a C-H bond. But in this case the C-H bond is in the form of an aldehyde, which comprises an oxidisable carbon (Figure 2.2). This aldehyde is oxidised to an acid and NADH + H^+ is produced (repeating theme: NAD and FAD specificity). But the acid formed is not simply succinic acid. It is the substituted acid succinyl CoA; this is the new concept (the energetics of an aldehyde oxidation), which is discussed next.

    2   An aldehyde oxidation is similar to a CHOH oxidation in that NAD is the electron acceptor. But the $\Delta_r G°$ of the reactants and products of the aldehyde reaction are such that (in simple terms) sufficient energy is released during the oxidation to enable some to be conserved as NADH + H^+, and some ('extra energy') to be conserved as the thioester succinyl CoA [3] (which we know from our discussion of acetyl CoA will have a hydrolysis reaction with a large K_{eq}). The hydrolysis of succinyl CoA is used in the next reaction, 5, which is a combination reaction (see concept: using ATP hydrolysis as part of a combination reaction with a large and negative ΔG), to 'drive' the production of GTP.[4] So one can look at a combination of reactions 4 and 5 as a reaction that (a) gets rid of the second (and final) carbon as CO_2, (b) prepares α-ketoglutarate for oxidation, (c) oxidises the second reduced bond, (d) oxidises an aldehyde to an acid, and (e) conserves the extra energy released from the aldehyde oxidation as ATP. The ATP produced in this reaction is quantitatively trivial (about 10%) in terms of the ATP equivalents that are produced in the form of reduced electron carriers by one turn of the Krebs cycle. But do not be misled by the numbers; this is a crucial reaction type. As will be seen when we look at glycolysis, if it were not for this characteristic of an aldehyde oxidation, it would not be possible for your cells to produce ATP in the absence of oxygen.

- The hard part of the Krebs cycle is now over. Numerically, carbon input equals carbon output; all that is left to do is to oxidise two more reduced bonds. The rest of the cycle is straightforward and follows exactly an alternating preparation/oxidation scheme. Succinate is an alkane, so can be immediately oxidised to the alkene fumarate in reaction 6. The alkane is one of the oxidisable substrates shown in Figure 2.2, and following the repeating theme: NAD and FAD specificity, FAD is the electron carrier. Note at this point that one of the C-H bonds in the alkane group of succinate is the one conserved when a valuable C-C bond was broken during the decarboxylation in reaction 3. Fumarate does not contain an oxidisable carbon so is rearranged in a preparation reaction (reaction 7) to produce malate, which contains a CHOH group (repeating theme: preparation reactions). This group is oxidised in reaction 8, NAD is the electron acceptor and the product is oxaloacetate which will accept the new acetyl group as the cycle begins another round.

Repeating exactly a sentence used at the beginning of this section on the Krebs cycle: 'The end result, the oxidation of four reduced bonds, and the production of four reduced electron carriers, two molecules of CO_2, and the original 4-carbon acceptor of the acetyl group'. Look back at the reactants and products of reactions 1 and 2 in Figure 2.2; these reactions will now mean a lot more to you.

We are now in a postion to consider the two pathways that produce the acetyl CoA for the Krebs cycle. Look back at Figure 3.1 and get your bearings. The electron transfer system, the Krebs cycle and acetyl CoA have been covered. The two remaining pathways are β-oxidation and glycolysis, which produce acetyl CoA from lipid and carbohydrate fuels respectively. β-oxidation will be addressed first as it is the simpler of the two.

4.3 β-oxidation: providing reduced bonds from lipid for the Krebs cycle

Figure 4.6 outlines the necessary details of the β-oxidation pathway. Again, all the details of the intermediates, the enzymes and the reactions are available in biochemistry texts and will not be covered here.

- The entry compound is a long-chain acyl CoA. The long-chain acyl group is a 16-carbon saturated fatty acid in Figure 4.6, with forty-six reduced bonds. Depending upon the cell and the organism, this starting compound can be longer or shorter, saturated or unsaturated, and odd or even numbered. You have effectively seen this compound before: it is acetyl CoA with the methyl group attached to the rest of the fatty acid molecule. The first two carbons will in fact become the acetyl CoA that enters the Krebs cycle. So the phosphopantetheine group is again carrying an acyl group, a longer one in this case (repeating theme: structural repeats: phosphopantetheine). Acyl CoA is also our first direct encounter with the concept of the combination reaction that involves ATP hydrolysis. The direct synthesis reaction for acyl CoA would have a positive ΔG. But the actual synthesis reaction in the cell includes ATP hydrolysis and therefore has a large and negative ΔG. The substrates are ATP, an acyl group and CoA (concept: using ATP hydrolysis as part of a combination reaction with a large and negative ΔG). The other lessons to be learned from an acyl CoA have already been discussed in the section on the Krebs cycle.

- As already mentioned, the rest of the pathway is very straightforward; it is simply a series of preparation and oxidation reactions (repeating theme: preparation reactions). But the interesting point here, and one that makes this pathway so easy to understand, is that it is an exact replica of a part of the Krebs cycle. This is a new and remarkable repeating theme (repeating theme: pathway repeat). Look back to Figure 4.4, and start at succinate. You will see an alkane to an alkene oxidation, followed by a hydration preparation reaction to produce a CHOH group, followed by another oxidation. The β carbon of succinate is oxidised from a CH_2 to a ketone, i.e. both of the reduced bonds have been oxidised and the electrons transferred to the electron carriers. This is exactly what happens in the first three reactions in the β-oxidation pathway, except that the 'succinate' in this case has a long saturated tail attached. This is the best example of a repeat so far – almost an entire pathway that we can simply skip over because we have covered all its aspects already. But note that not all of the reduced bonds of lipid fuels end up as acetyl CoA: each 'cycle' of β-oxidation produces four reduced bonds as acetyl CoA and two as reduced electron carriers. The latter, of course, bypass the Krebs cycle and enter the electron tranfer system.

- Reaction 4 in β-oxidation splits off acetyl CoA and uses another CoA to produce an acyl CoA that is two carbons shorter. This acyl CoA will be used as a substrate for reaction 1 in another β-oxidation 'cycle' in which the next β carbon (original carbon # 5) will be oxidised.

- Remember recycling (repeating theme: recycling). Note that the CoA reactant in reaction 4 can be considered to be recycled from an acetyl CoA product that has donated its acetyl group to the Krebs cycle.

- Also remember the value of C-C bonds, and the fact that they cannot be oxidized directly. In the Krebs cycle we saw two examples of C-C bonds being conserved as C-H bonds, and eventually oxidised. There is another example of this in reaction 4 of β-oxidation. The C-C bond between the α and β carbons of the reactant is cleaved and conserved as a C-H bond in the methyl group of the acetyl CoA product (repeating theme: preparation reactions). This bond will be oxidised in the Krebs cycle.

The entire process by which the lipid fuels are oxidised has now been covered, and one example of the involvement of ATP hydrolysis in work (the synthesis of acyl CoA) has been encountered. So in terms of the overall metabolic scheme, we are almost finished with the right-hand side of Figure I.1. What we have not covered is the initial preparation pathway for carbohydrate oxidation. This pathway will not, as you might predict, offer anything new in terms of concepts or themes. But glycolysis offers more examples of ATP hydrolysis and work, and re-emphasises the role of the K_{eq} of reactions. But above all, it offers some exquisite reiterations, of one concept and one theme in particular, which are integral to a most fascinating aspect of glycolysis, i.e. its function in producing ATP in the absence of oxygen.

4.4 Glycolysis: multi-purpose processing

Glycolysis is a complex entity. It is difficult to simplify for the very reason that it is multi-purpose. On the one hand it does involve a straightforward oxidation, but on the other, the number of reduced bonds in the products is the same as that in the substrate. It can also be considered as a large preparation reaction, as it can produce from carbohydrate, the prerequisite for acetyl CoA, the starting point for fuel oxidation. But it contains reactions that fulfil a metabolic control function, and it supplies carbon and reduced bonds to the pentose phosphate pathway (a pathway that is not covered in this book), which plays a role in the synthesis and the recycling of nucleic acid bases. So we will start with a simplified glycolytic scheme, in order to explain the aspects of glycolysis that are relevant to energy extraction, and which reiterate important concepts and themes. We will eventually look at all the reactions, but only when the role of glycolysis in energy metabolism is clear.

Start with Figure 4.7. There are two questions that can be identified in this figure, and which are discussed below.

- First, how does glycolysis function to provide reduced bonds for the Krebs cycle when glucose is being completely oxidised? There are two issues to be discussed within this question.

1. Glucose has six carbons and twelve reduced bonds (see Figure 1.2). The reactions in glycolysis produce two moles of pyruvate which has three carbons and five reduced bonds. So ten of the twelve reduced bonds of glucose appear in pyruvate, which can then be converted to acetyl CoA (see point 2 below). The other two reduced bonds appear as $NADH + H^+$, which poses a cycling problem, one that was anticipated in section 4.2 (repeating theme: recycling). The quantities of NAD in the cell are very small when considered in the context of the continuous flux of carbon through glycolysis that is required, for example, by your brain, which uses only glucose. In the cells in the brain, there is only enough NAD for three minutes at resting metabolic rate. So it is imperative that the $NADH + H^+$ produced by glycolysis (Figure 4.7) be continually recycled so that NAD is continually available as an electron acceptor. When glucose is being completely oxidised, this recycling is accomplished by a hydrogen shuttle (Figure 4.7). The reactions involved in the various forms of hydrogen shuttles can be found in biochemistry texts and will not be discussed here. But in the context of this book, the most significant attribute of a hydrogen shuttle is that the electrons are shuttled into the mitochondria, where they can enter the ETS and eventually become part of a water molecule. Thus the principle of innocuous end-products is maintained (concept: innocuous end-products), which is an essential component of a long-term oxidative process.

2. The pyruvate is converted to acetyl CoA in a reaction catalysed by pyruvate dehydrogenase (Figure 4.8, reaction 1). This represents nothing new, which is why it is interesting. It is a new repeating theme (repeating theme: reaction repeats: aldehyde oxidation), and is a virtual replica of reaction 4 in the Krebs cycle (Figure 4.4 and Figure 4.8, reaction 2). The 'extra energy' from the aldehyde oxidation is conserved as a thioester bond in acetyl CoA (vs succinyl CoA in reaction 4 of the Krebs cycle; concept: the energetics of an aldehyde oxidation).

So the culmination of glycolysis in the presence of oxygen is that one of the five reduced bonds of pyruvate appears as $NADH + H^+$ and enters the ETS, while the other four bonds enter the Krebs cycle as an acetyl group.

- Second, how does glycolysis function to produce ATP in the absence of oxygen? This question is best answered by considering the recycling issue first (repeating theme: recycling).

4: OXIDATIVE METABOLISM

1. In the absence of oxygen, the NADH + H$^+$ cannot be recycled using a hydrogen shuttle, as a hydrogen shuttle requires oxygen (Figure 4.7). The recycling is accomplished by reducing pyruvate to lactate (Figure 4.7 and Figure 4.9). This is an interesting variation on a familiar theme. You have seen similar reactions before, but in reverse. For example, look at the first reaction in Figure 2.2, and reactions 3 and 8 in the Krebs cycle (Figure 4.4). The difference here is that this is a reduction instead of an oxidation, the ketone group is being reduced to a CHOH group, and consequently NADH + H$^+$ is being oxidised to NAD. So the recycling problem is solved, but the resulting product, lactate, produces its own problems.

   Lactate is not an innocuous end-product (concept: innocuous end-products). Lactate cannot accumulate like water, and cannot be excreted like CO_2. The accumulation of lactate inhibits and limits glycolysis. This is one reason why anaerobic muscle contraction is limited in terms of time. Marathon runners (aerobic muscle contraction) only have to stop when they run out of fuel at two to three hours, whereas a 100 m sprinter or a cheetah have only seconds before the inhibition of glycolysis manifests as a decrease in speed.

2. The production of lactate, which has 6 reduced bonds, also raises the most interesting question in glycolysis. If glycolysis in the absence of oxygen begins with one mole of glucose (twelve reduced bonds), and ends with two moles of lactate (twelve reduced bonds), how can it produce any ATP? This issue was mentioned during the discussion of the oxidation of α-ketoglutarate in the Krebs cycle. There is an oxidation of an aldehyde to an acid in glycolysis (Figure 4.7). But the NADH + H$^+$ produced cannot be a source of ATP because as mentioned above, the ETS is not available. The only ATP produced by glycolysis is generated from that 'extra energy' that arises from an aldehyde oxidation. This is why the concept involving aldehyde oxidation (the energetics of an aldehyde oxidation) is crucial for glycolysis. Without this 'extra energy', a cell would not be able to make ATP in the absence of oxygen. 100 metre sprints would not exist, as this type of intense exercise requires a rate of ATP production that cannot be matched by oxygen delivery.

   The reaction involved is an interesting one (Figure 4.8 reaction 3; Figure 4.9, reactions 6 and 7), as it is a variation on the theme demonstrated by the two other aldehyde oxidations we have encountered, the oxidation of pyruvate and α-ketoglutarate (repeating theme: reaction repeats: aldehyde oxidation). In the case of the aldehyde oxidation in glycolysis, (a) the initial substrate is an aldehyde, so a decarboxylation is not necessary, and (b) the 'extra energy' from the aldehyde oxidation is conserved as a phosphate anhydride bond (the same structure as the terminal bond in ATP, so it is a structural repeat: phosphate anhydride bond; Figure 2.6), instead of a thioester bond. This phosphorylated intermediate is then hydrolysed as part of a combination reaction that synthesises ATP. This is in one sense a repeat of the reaction catalysing the hydrolysis of succinyl CoA in the Krebs cycle, which is a combination reaction (see concept: using ATP hydrolysis as part of a combination reaction with a large and negative ΔG), that 'drives' the production of GTP.

This reaction also provides the opportunity to introduce another structural repeat, that involving sulphydryl groups (repeating theme: structural repeats: sulphydryl groups). Think back to the discussion of acetyl CoA (section 4.2). The CoA contained a sulphydryl group on the phosphopantetheine component of the molecule, which bound the acyl group of the acetyl CoA, and that of the long-chain acyl CoA molecules. The complex aldehyde oxidation reactions also involve an acyl intermediate. The enzymes that oxidise pyruvate and α-ketoglutarate co-opt lipoic acid as a prosthetic group (repeating theme: prosthetic groups). The lipoic acid contains a lipoyl group (which in turn contains SH groups) which binds the acyl group intermediate. The enzyme that oxidises the aldehyde in glycolysis contains a cysteine which contains an acyl-binding SH group in its R group.

The detailed pathway is outlined in Figure 4.9 but does not contain any other reactions that bear directly on the principles forming the basis of this book. It will, however, be briefly considered in order to put all the reactions in the context of the discussions above, and to prevent any confusion that might arise from some of the reactions.

- The first three steps accomplish two phosphorylations, which have no implications for our discussion. But they are instructive in that each reaction that uses ATP as a substrate is an example of a combination reaction. In this case the purpose of the ATP is not to drive an unfavourable reaction (see concept: using ATP hydrolysis as part of a combination reaction with a large and negative ΔG), it is to produce a reaction with a high K_{eq} (concept: using ATP hydrolysis to increase the K_{eq} of a reaction). This is exactly what will happen if ATP is incorporated as a substrate into a reaction that previously had a low (approximately 1.0) K_{eq} (see sections

1.1–1.3 and 2.3). Reactions with a high K_{eq} are used as controlling steps in pathways, for the very reason that they are insensitive to the concentrations of their reactants and products. Reactions 1 and 3 are invested with most of the control over the rate of glycolysis. Note in contrast that reaction 2 (an isomerisation), in which the total energy of reactants and products (at equal concentrations) would be virtually equal, has a K_{eq} close to 1.0 and exerts very little control over the rate of the pathway, and merely responds to the concentrations of its reactants and products.

- The fructose 1,6-bisphosphate is then split into two 3-carbon compounds. Note that only one of them is an aldehyde (the dihydroxyacetone phosphate can be used for other purposes), but the two are interconvertible. So for our purposes it is easiest to consider that the bisphosphate is split into two aldehydes (Figure 4.9, reaction # 5).

- The next two steps (reactions 6 and 7) comprise the aldehyde oxidation and the production of ATP that have been discussed in detail above. In the context of this book, these steps can be considered as those that produce the ATP in the absence of oxygen.

- The next two steps (reactions 8 and 9) result in the production of an enol phosphate. Again this compound has no direct bearing on our discussion, but is interesting in that its hydrolysis reaction has a higher K_{eq} that that of the hydrolysis of ATP. This attribute is therefore used in the next step to phosphorylate ADP. For our purposes, reaction 10 can be considered as the step (remember there are two moles of phosphoenol pyruvate produced per mole of glucose) that 'pays back' the two moles of ATP used earlier in the pathway.

- The final reaction (11) has a recycling function and has already been discussed in detail.

4.5 Energy content: carbohydrate vs lipid

Armed with a detailed knowledge and comprehensive understanding of oxidative metabolism, it is now an opportune time to return to Figure 1.1, and express the energy content of the fuels as something more tangible than reduced bonds, i.e. as ATP. This necessarily involves two issues; one is controversial, the other takes us back to redox potentials. The controversial one is that it is not really known how many moles of ATP are produced by the ETS when one mole of NADH + H^+, or $FADH_2$ are oxidised. So the most recent estimations, which are the most accurate, have been used.

The other issue, which was not mentioned during the discussion of the ETS, is that more ATP are produced from the oxidation of NADH + H^+, compared to that of $FADH_2$. The reason for this is that FAD has a higher affinity for electrons than NAD, which is why in Figure 4.1, FADH$_2$ donates electrons to Complex II, rather than to Complex I as does NADH + H$^+$. Since the transfer of electrons between Complex I and II involves proton pumping, it follows that the oxidation of FADH$_2$ will involve less proton pumping than the oxidation of NADH + H$^+$ and therefore less ATP synthesis. So all reduced bonds are not created equal in terms of potential energy yield upon oxidation. In fact we could now rank them in descending order, from those in an aldehyde, to those in a CHOH, to those in an alkane. For the purposes of this discussion it will be assumed that 2.5 and 1.5 moles of ATP respectively are synthesised per mole of NADH + H$^+$ and FADH$_2$ oxidised.

These numbers have been used to produce the energy yields in Figure 4.10, which demonstrates several interesting points. It is timely to remember that these points can only now be appreciated because of your understanding of the concepts and themes that are the basis of the process of energy extraction in the cell.

- For each of the fuels, the number of reduced electron carriers produced equals the number of reduced bonds in the molecule.

- The 6-carbon lipid is more reduced than the 6-carbon carbohydrate. This manifests as a higher production of ATP from the lipid, in terms of total/molecule, ATP/g and ATP/carbon. The difference in reduction is reflected in the greater number of C-O bonds in the carbohydrate compared to the lipid (see Figure 4.10 and the general formulae for the two types of fuel in Figure 1.1).

- The slightly lower ATP production per reduced bond from the lipid reflects the differential ATP yield of FADH$_2$ vs NADH + H$^+$ that was discussed above. There are more alkane oxidations (which produce FADH$_2$) associated with lipid oxidation (in β-oxidation Figure 4.6) due to the nature of the molecule, which is essentially a long alkane (see repeating theme: NAD and FAD specificity).

That completes the treatment of energy extraction, or oxidation. Chapters 1–4 have introduced the concepts and themes involved, and shown how the four major metabolic pathways can be considered as sequential components of an overall, relatively simple scheme. But what about the reduction part of Figure I.1? We have already encountered one example of the synthesis of reduced bonds, the reduction of pyruvate to lactate as a recycling strategy in glycolysis in the absence of oxygen. In the next chapter, reduction in animal cells will be considered in more detail, in the form of the pathway of fatty acid synthesis.

affinity for electrons than NAD, which is why in Figure 4.1, $FADH_2$ donates electrons to Complex II, rather than to Complex I as does $NADH + H^+$. Since the transfer of electrons between Complex I and II involves proton pumping, it follows that the oxidation of $FADH_2$ will involve less proton pumping than the oxidation of $NADH + H^+$ and therefore less ATP synthesis. So all reduced bonds are not created equal in terms of potential energy yield upon oxidation. In fact we could now rank them in descending order, from those in an aldehyde, to those in a CHOH, to those in an alkane. For the purposes of this discussion it will be assumed that 2.5 and 1.5 moles of ATP respectively are synthesised per mole of $NADH + H^+$ and $FADH_2$ oxidised.

These numbers have been used to produce the energy yields in Figure 4.10, which demonstrates several interesting points. It is timely to remember that these points can only now be appreciated because of your understanding of the concepts and themes that are the basis of the process of energy extraction in the cell.

- For each of the fuels, the number of reduced electron carriers produced equals the number of reduced bonds in the molecule.

- The 6-carbon lipid is more reduced than the 6-carbon carbohydrate. This manifests as a higher production of ATP from the lipid, in terms of total/molecule, ATP/g and ATP/carbon. The difference in reduction is reflected in the greater number of C-O bonds in the carbohydrate compared to the lipid (see Figure 4.10 and the general formulae for the two types of fuel in Figure 1.1).

- The slightly lower ATP production per reduced bond from the lipid reflects the differential ATP yield of $FADH_2$ vs $NADH + H^+$ that was discussed above. There are more alkane oxidations (which produce $FADH_2$) associated with lipid oxidation (in β-oxidation Figure 4.6) due to the nature of the molecule, which is essentially a long alkane (see repeating theme: NAD and FAD specificity).

That completes the treatment of energy extraction, or oxidation. Chapters 1–4 have introduced the concepts and themes involved, and shown how the four major metabolic pathways can be considered as sequential components of an overall, relatively simple scheme. But what about the reduction part of Figure I.1? We have already encountered one example of the synthesis of reduced bonds, the reduction of pyruvate to lactate as a recycling strategy in glycolysis in the absence of oxygen. In the next chapter, reduction in animal cells will be considered in more detail, in the form of the pathway of fatty acid synthesis.

1.1–1.3 and 2.3). Reactions with a high K_{eq} are used as controlling steps in pathways, for the very reason that they are insensitive to the concentrations of their reactants and products. Reactions 1 and 3 are invested with most of the control over the rate of glycolysis. Note in contrast that reaction 2 (an isomerisation), in which the total energy of reactants and products (at equal concentrations) would be virtually equal, has a K_{eq} close to 1.0 and exerts very little control over the rate of the pathway, and merely responds to the concentrations of its reactants and products.

- The fructose 1,6-bisphosphate is then split into two 3-carbon compounds. Note that only one of them is an aldehyde (the dihydroxyacetone phosphate can be used for other purposes), but the two are interconvertible. So for our purposes it is easiest to consider that the bisphosphate is split into two aldehydes (Figure 4.9, reaction # 5).

- The next two steps (reactions 6 and 7) comprise the aldehyde oxidation and the production of ATP that have been discussed in detail above. In the context of this book, these steps can be considered as those that produce the ATP in the absence of oxygen.

- The next two steps (reactions 8 and 9) result in the production of an enol phosphate. Again this compound has no direct bearing on our discussion, but is interesting in that its hydrolysis reaction has a higher K_{eq} that that of the hydrolysis of ATP. This attribute is therefore used in the next step to phosphorylate ADP. For our purposes, reaction 10 can be considered as the step (remember there are two moles of phosphoenol pyruvate produced per mole of glucose) that 'pays back' the two moles of ATP used earlier in the pathway.

- The final reaction (11) has a recycling function and has already been discussed in detail.

4.5 Energy content: carbohydrate vs lipid

Armed with a detailed knowledge and comprehensive understanding of oxidative metabolism, it is now an opportune time to return to Figure 1.1, and express the energy content of the fuels as something more tangible than reduced bonds, i.e. as ATP. This necessarily involves two issues; one is controversial, the other takes us back to redox potentials. The controversial one is that it is not really known how many moles of ATP are produced by the ETS when one mole of NADH + H^+, or $FADH_2$ are oxidised. So the most recent estimations, which are the most accurate, have been used.

The other issue, which was not mentioned during the discussion of the ETS, is that more ATP are produced from the oxidation of NADH + H^+, compared to that of $FADH_2$. The reason for this is that FAD has a higher

Notes

[1] We are proceeding in steps towards conserving the energy from reduced bonds as a maintenance of a non-equilibrium ratio of ATP and its hydrolysis products (see Chapter 2, note 4). So far the energy has been conserved first as reduced electron carriers, now as a proton gradient.

[2] Figure 4.4 shows the ketone carbon of oxaloacetate to be one of the carbons that was incorporated from the previous acetyl group. So it is timely at this point for you to realise that none of the reduced bonds or carbons from the incoming acetyl group are processed in the first round of the cycle. It actually takes three turns of the cycle to process the incoming acetyl group. This is not an important concept or theme, and is only mentioned at this stage in case it causes confusion.

[3] Look back at Chapter 2, note 4, and make sure you realise that our language is still not absolutely correct in thermodynamic terms.

[4] GTP is thermodynamically equivalent to ATP and is used in some reactions as an energy source instead of ATP. For the purposes of our discussion of the Krebs cycle you should consider the GTP as ATP.

5 Synthesising Reduced Bonds in Animal Cells: Fatty Acid Synthesis

Look back at section 1.5. You oxidise a large number of reduced bonds each day, about 1.4×10^{22} of them. We are now familiar with the form in which the reduced bonds occur, and understand how the energy they contain is conserved as ATP. But where do all these reduced bonds ultimately come from, how is the oxidative process reversed to synthesise them, and where does this happen?

These reductive processes occur in both plant and animal cells, but are quantitatively much more significant in the former. A reductive pathway in animal cells (fatty acid synthesis) will be considered first. This will emphasise four key points (listed below) about reductive pathways in animal cells that will set the scene for the following chapter on photosynthesis.

1. The reactions that comprise the reductive pathways in animals are instructive as they conform to, and widely reiterate, the concepts and themes that have been catalogued in the previous four chapters.

2. The reactions that catalyse the formation of C-H bonds are essentially the oxidation reactions in reverse, and the electrons derive from reduced electron carriers.

3. The formation of C-C bonds is 'driven' by the hydrolysis of ATP.

4. The starting compound for reduction pathways in animal cells already represents almost 70% of the reduction that will be found in the final product. These pathways in animal cells therefore accomplish very little net reduction.

An enzyme complex (the fatty acid synthase complex) is responsible for the synthesis of fatty acids in animal cells. The exact nature of the complex differs between yeast and vertebrates for example, but the structural principles are the same, i.e. the complex comprises binding groups, about six enzymes and various prosthetic groups. Fatty acid synthesis is a complicated process as you will see shortly, and as with glycolysis, one can become distracted from the salient points by the complexity. But the complexity is only in the hardware; the concepts and themes will all be familiar. As was the case with glycolysis, a simplified scheme will be initially considered (Figure 5.1), to enable you to visualise the entire process, before it is considered in detail. This figure describes the process by which a 6-carbon saturated fatty acid (see Figure 4.10) is synthesised. The figure will of necessity simplify some of the steps to the point of error, but it remains faithful to the concepts involved, and the reactions will be discussed in detail subsequently.

5.1 Fatty acid synthesis: the basic strategy

Look at Figure 5.1, which is fully explained in its legend. Go through this figure carefully before proceeding; it is necessary to be familiar with the process in order to appreciate the details in the next figure. The underlying strategy of the synthesis of long-chain saturated fatty acids is carbon entry, followed by carbon condensation to produce a longer chain, followed in turn by reduction of any carbons that are not fully reduced. This pattern simply repeats until the chain is of the right length (Figure 5.1). Note three points about this process.

1. It synthesises C-C bonds, in reactions 3 and 7. These condensations are 'driven' by ATP hydrolysis, in an indirect way, as will be discussed shortly. These syntheses are not however net reductions, as will also be discussed shortly.

2. It synthesises C-H bonds, when carbons 5 and 3 are reduced in reactions 4 and 8. The electrons come from reduced electron carriers.

3. It therefore synthesises two of the reduced bonds that comprise the energy in a fuel molecule that is available to a cell. But it is already obvious from this figure that the amount of reduction that occurs in this process (the partial reduction of only two of the six carbons) is a small proportion of that found in the hexanoate in Figure 4.10. This is the first hint of the significance of the major reducing process in the biosphere, photosynthesis.

5.2 Fatty acid synthesis: recurring concepts and themes, and a final message

The detailed pathway is shown in Figure 5.2. As previously mentioned, the process is complex, and there are distractions. But the process has to be described in order to appreciate the concepts and themes.

- The entry carbons are in a familiar form, acetyl CoA, which was discussed in detail in section 4.2. As noted in Figure 5.1, the first two carbons of each fatty acid (which will become the methyl end of the fatty acid) are added to a binding protein on the fatty acid synthase (FAS) complex (reaction 2). This binding protein is also the enzyme that will

5: SYNTHESISING REDUCED BONDS IN ANIMAL CELLS

catalyse the carbon condensation, and is called the condensing enzyme. It binds acyl groups using a sulphydryl group on a specific cysteine residue. This is a repeat of the strategy used by the enzyme involved in the aldehyde oxidation in glycolysis (section 4.4; repeating theme: structural repeats: sulphydryl groups).

- All other carbons from this point are added as a 3-carbon compound (malonyl CoA; see Figure 5.2, reaction 3). Malonyl CoA is bound to acyl carrier protein (Figure 5.2), which like CoA, incorporates the non-protein pantothenic acid into its structure in order to bind the acyl moiety of malonyl CoA; another structural repeat (repeating theme: structural repeat: phosphopantetheine; repeating theme: prosthetic groups).

- Malonyl CoA is synthesised from acetyl CoA, and the reaction (1) raises three interesting points.

    1 Although this reaction results in the synthesis of a C-C bond (see * on Figure 5.2), there is not a net synthesis of reduced bonds, as this C-C bond can be considered as a conservation of the C-H bond which is lost from the methyl group of the acetyl CoA reactant. This is in keeping with the fact that reduced bonds are precious, and so are always conserved. This point also represents another hint of the impact of photosynthesis on net reduction, as this C-C bond was already effectively present in the acetyl CoA.

    2 The involvement of ATP hydrolysis (see concept: using ATP hydrolysis to increase the K_{eq} of a reaction), as in the initial reactions of glycolysis, ensures that the reaction has a high K_{eq} and therefore will proceed under all conditions, and not be governed by the concentrations of its reactants and products.

    3 The incorporation of CO_2 is only temporary – it is lost in a later reaction – but the C-C bond is conserved (see reaction 4 below). So the only carbons that are actually incorporated into the growing fatty acid are those from acetyl CoA.

- The next reaction (4) is the condensation reaction that occurs with each 2-carbon 'cycle' of fatty acid synthesis, and is the reaction that is responsible for lengthening the chain. The carbonyl carbon (see arrow on Figure 5.2) of acetyl CoA is condensed with the methylene carbon of malonyl CoA, and this reaction again raises three points.

    1 As in point 1 above, the C-C bond formed in the 4-carbon product is not a net synthesis of a reduced bond. This reduced bond is essentially the C-H bond that was lost from the methyl group of the acetyl CoA

5: SYNTHESISING REDUCED BONDS IN ANIMAL CELLS

reactant during the synthesis of malonyl CoA. This re-emphasises an underlying theme of this chapter, which is how little reduction is accomplished in animal cells.

2. The effect of the incorporation of ATP hydrolysis into the reaction that synthesises malonyl CoA, is effectively carried over into this reaction, i.e. this reaction is 'driven' by the high K_{eq} of the reaction that produces one of its reactants.

3. The product of this reaction now contains the β-carbon that will be the target of reduction in the next three reactions.

- The next three reactions (5, 6 and 7) are catalysed by enzymes that are part of the synthase complex, and accomplish the oxidation of the β-carbon of the 4-carbon compound that is attached to acyl carrier protein. These reactions comprise an interesting repeat, but also set the scene for the introduction of another underlying theme of this chapter: reductive pathways in animal cells are not new pathways, they are oxidation pathways in reverse.

1. These three reactions are β-oxidation and the Krebs cycle revisited, but in reverse (repeating theme: pathway repeat). Instead of β-oxidation, this is β-reduction. Look back at Figure 4.4 and Figure 4.6. Find the three β-oxidation reactions in both of these pathways and run them in reverse, starting with the β-carbon ketone. You will find that the resulting sequence is essentially that of reactions 5, 6 and 7 in Figure 5.2. (Revisit the reactions in Figure 2.2, but remember to look at them in reverse.) Reaction 5 is a ketone to a CHOH reduction. Reaction 6 is a preparation reaction (see repeating theme: preparation reactions) as a CHOH group is not a reactant for a reduction reaction (see Figure 2.2), and reaction 7 is an alkene to an alkane reduction. So reduction in animal cells really is a straightforward reversal of oxidation.

2. There are however, two differences in detail between the oxidative and reductive pathways. First, NADP is used as the electron donor. This is a new compound, but the structural difference between it and NAD was mentioned in Figure 2.3. There is nothing new about NADP; the redox potentials of the NAD/NADH + H^+ and NADP/NADPH + H^+ pair are identical. NADP is probably used in the synthetic pathway in order to distinguish between oxidation and reduction for regulatory purposes. NADPH + H^+ is the product of various oxidation reactions that are similar to those that have already been discussed, and will not be considered further. Second, NAD is used for both the ketone/CHOH and alkane/alkene transition, FAD

plays no role in this synthetic pathway. This contradicts the rule concerning NAD and FAD specificity that was introduced in Figure 2.3, and to my knowledge there is no explanation for this aberrance.

- The 4-carbon reduced acyl product of reaction 7 is then moved to the cysteine sulphydryl of the condensing enzyme, another malonyl CoA binds to the ACP and the condensation reaction (4) produces a 6-carbon compound with a new β-carbon ketone, which is then reduced. The cycles of reduction can continue until a long-chain (e.g. sixteen carbons) saturated fatty acid is produced and cleaved off the ACP. But for our purposes a 6-carbon chain is long enough to demonstrate the relevant points, and is shown as the final product in Figure 5.2 (reaction 8).

- Look at the 6-carbon product of reaction 8. Count the reduced bonds, there are sixteen, as was already stated in Figure 4.10, but only four of these (marked in Figure 5.2) were synthesised by the reductive process described. So only 25% of the reduced bonds in the fatty acid fuels in your cells are actually synthesised 'in-house'. The rest were already present in the acetyl CoA building blocks. The situation is slightly more complex for carbohydrate synthesis in animal cells, but the result is similar. Only about 17% of the reduced bonds in glucose are synthesised 'in-house' by the carbohydrate synthesis pathway in animals called gluconeogenesis.

The 'in-house' reduction issue returns us to the fourth point on p.40, and raises a question. What is the origin of the other 75–80% of the reduced bonds in the fuels in animal cells? Obviously the system in Figure I.1 cannot work without the reduction side of the equation, and reduction in animal cells is not the answer. The answer, of course, is photosynthesis, and it is to this reduction specialist that we next turn our attention.

6 Synthesising Reduced Bonds in Plant Cells: the Calvin Cycle and Photo-reduction

The process of photosynthesis uses light energy to produce ATP and NADPH + H$^+$, which are then used as a source of energy and reducing power to synthesise reduced bonds in the form of glucose. As stated at the end of Chapter 5, this glucose represents 75–80% of the reduced bonds that are found in the fuels used by animal cells. There are three aspects of photosynthesis that are of interest in the context of this book. These are:

1 the reactions involved in the synthesis of glucose;
2 the synthesis of the reducing power required for the reactions in 1 (above); and
3 the synthesis of ATP, which is also required for the reactions in 1 (above).

These will be discussed in turn, discussions that will again comprise mostly repeats of concepts and themes. But these discussions will also introduce a new concept that could well compete with the mitochondrial electron transfer system for the title of the most important energy transduction mechanism in the biosphere.

This section will conclude the analysis of the reduction side of Figure I.1, and render the oxidation/reduction cycle complete. The only concept that will remain to be explained in Figure I.1, which is outside the scope of this book, is the basis of the ultimate source of power for the scheme depicted in Figure I.1. The reactions involved, which occur in the sun, generate 10^{27} joules per second and are accompanied by the loss of 4×10^6 tonnes of mass per second.

6.1 The reactions involved in the synthesis of glucose: the Calvin cycle

The Calvin cycle is difficult to portray in a diagram, as its stoichiometry is such that multiple reactions must occur simultaneously, and the same compound must be processed through different reactions simultaneously, if any sense is to be made of a final overall reaction.

Figure 6.1.A shows the Calvin cycle as a list of reactions. It is explained in the legend, but take some time to satisfy yourself that the final reaction is a faithful representation of the fifteen different reactions. There are five issues associated with reactions 1–8 that need to be discussed.

1. The carbon input is CO_2, (reaction 1; see also Figure I.1 and Figure 2.1). But the important point, in terms of reduction, is that this reaction (Figure 6.1.B) does not represent a net synthesis of reduced bonds. Ribulose 1,5-diphosphate has ten reduced bonds, and the products (2 × 3-phosphoglycerate) represent ten reduced bonds. Neither is there is a change in the proportion of the different bonds, as ribulose 1,5-diphosphate has four C-C bonds and six C-H bonds, and 3-phosphoglycerate has two C-C bonds and three C-H bonds. As with fatty acid synthesis in animals (section 5.2), the incorporation of carbon does not represent a net synthesis of reduced bonds.

2. The next two reactions of Figure 6.1.A (2 and 3) represent the synthesis of the reduced bonds that is the driving force, courtesy of the sun, of the cycle represented in Figure I.1. However, seminal as these reactions are, there is no need to discuss them in great detail, as they are a virtual repeat, although in reverse, of the aldehyde oxidation reaction in glycolysis (concept: the energetics of an aldehyde oxidation; repeating theme: pathway repeat). Look back to Figure 4.8 (reaction 3) and Figure 4.9 (reactions 6 and 7) and refresh your memory of this oxidation reaction that produces $NADH + H^+$. Reactions 2 and 3 in the Calvin cycle comprise the same sequence, but in reverse, and use NADP instead of NAD as the electron acceptor/donor. So this is an acid/aldehyde reduction, the reverse of the aldehyde oxidation reaction shown in Figure 2.2, and, take note, represents the *only*, and *all* the net reduction that takes place as a result of photosynthesis. Also note that this concept and pathway repeat is occurring across two metabolic systems and two different kingdoms.

3. Reactions 2 and 3 of the Calvin cycle comprise the reduction of an acid to an aldehyde. So it follows that since the most significant character of an aldehyde oxidation was the 'extra energy' produced (concept: the energetics of an aldehyde oxidation), 'extra energy' is required as a reactant for the reverse reaction. This can be seen as the ATP reactant in reaction 2. So the reduction in photosynthesis requires $NADPH + H^+$ and ATP. ATP synthesis is addressed in section 6.3.

4. Reactions 4 to 8 effectively convert the twelve reduced bonds that have been synthesised, into glucose, the compound that has represented the carbohydrate fuel throughout this book. With two exceptions (discussed in point 5 below), these reactions are reversals of the respective reactions of glycolysis shown in Figure 4.9. This reflects the fact that the

concentrations of the reactants and products of these reactions are such that these reactions are close to equilibrium in the cell (they have low K_{eq}s and low ΔGs). The reactions can therefore proceed in either direction, depending upon the reactant and product concentrations, as was the case for phosphoglucoseisomerase in glycolysis (Figure 4.9).

But reactions 1 and 3 in Figure 4.9 are different. They are 'driven' by ATP hydrolysis in the glycolytic direction. Given the oft-repeated concept of ATP hydrolysis 'driving' reactions (concept: using ATP hydrolysis as part of a combination reaction with a large and negative ΔG), it seems contradictory that these reactions can be reversed within the constraints of cell structure and function (remember that the ΔG for the reaction must be negative to proceed, regardless of the direction in which it is proceeding). And so it is, as explained below.

5   The explanation is that reactions 6 and 8 in Figure 6.1.A, are not the exact reverse of the respective reactions in glycolysis (reactions 1 and 3 in Figure 4.9). Note that reactions 6 and 8 in Figure 6.1.A do not involve the synthesis of the phosphate anhydride bond of ATP. Instead, the phosphate group from the phosphate ester is simply released as inorganic phosphate. These reactions 'bypass' the problem of reversing ATP hydrolysis. Both of these reactions have low K_{eq}s, they are catalysed by different enzymes from those that catalyse the reverse reactions in glycolysis, and demonstrate that indeed, without some sort of energy input it is impossible to drive the hydrolysis of ATP backwards within the concentration constraints of the cell.

Reactions 9 to 15 resynthesise the initial reactant for reaction 1, ribulose 1,5-diphosphate and will not be discussed further. Reaction 15 requires ATP, the synthesis of which is addressed in section 6.3.

6.2 The synthesis of the NADPH + H$^+$ required for the reduction in the Calvin cycle

The concept of redox potential (concept: redox potential) resurfaces in the discussion of this issue. Remember that a redox reaction is like any other reaction and must have a negative ΔG to proceed. As discussed in Section 2.2, a negative ΔG can be a result of favourable redox potentials of the electron donor and acceptor. Or, unfavourable potentials can be overcome (to some extent) by the concentrations of reactants and products. In the case of the light-fuelled synthesis of NADPH + H$^+$, the strategy is one of favourable redox potentials. How this is achieved is the essence of photosynthesis, and comprises the final and arguably the most significant concept addressed in this book.

In summary, NADPH + H$^+$ is synthesised by reducing NADP with the electrons from water. So, as in the ETS in mitochondria, this process involves an electron transfer system, which transfers electrons from water to NADP. Figure 6.2.A is a diagram of a chloroplast, the organelle in plants in which photo-reduction takes place. It contains invaginations called thylakoids. In the thylakoid membranes are photosystems that comprise reaction centres (Figure 6.2.B) such as P680 and P700, and electron transfer components that transfer electrons from one reaction centre to another, or from P700 to NADP. So water is split in the lumen of the thylakoid, the electrons are passed to P680, then to P700 and finally to NADP to form NADPH + H$^+$. The details of this process, and the mechanisms involved, are still under investigation, but what is known is presented in detail in biochemistry texts.

The scheme looks simple enough, and involves electron transfer, which is a familiar concept. But there is a problem with this scheme which is outlined below.

- Look back at the E'_0 values in Figure 4.1.

- What Figure 6.2.B depicts is the transfer of electrons from the oxygen/water pair (E'_0 = 0.82 V) to the NADP/NADPH + H$^+$ pair (E'_0 = that of NAD/NADH + H$^+$ = -0.32). This is essentially a reversal of the scheme represented in Figure 4.1, which represents the mechanism by which about 90% of the ATP in animal cells is synthesised.

- But the donor in this case has a much higher affinity for electrons than the acceptor.

- The difference in affinities in this case is not only unfavourable, it is enormous. To put the differences into context, they are much larger than that which was overcome by reactant and product concentrations in Figure 2.5.

This problem is overcome by a mechanism that represents the new concept that was mentioned previously. Light energy is used to change the redox potentials of redox pairs (concept: light energy can change redox potential).

This mechanism is depicted in Figure 6.3. The detailed mechanisms involved in this scheme are complex, in some cases still not fully understood, and will not be addressed here.

- Light energy is initially used to split water into electrons, protons and oxygen. This occurs in the water-splitting enzyme complex (Figure 6.2) and is a separate issue from that involving light energy and the perturbation of redox potentials. Note that the reactants and products of this reaction are the water and oxygen depicted in Figure I.1 and Figure 2.1.

- The electrons are transferred, down a redox potential gradient, to the P680 reaction centre.
- Light energy is then harnessed to decrease the redox potential of the reduced/oxidised P680 redox pair; this is the new concept in action (light energy can change redox potential).
- The electrons are again transferred, through a series of electron acceptors and donators, down a redox potential gradient, to another reaction centre, P700. The redox potential of the P700 redox pair at this stage is approximately 0.4 V, still higher than that of the NADP/NADPH + H^+ pair (-0.32 V).
- So light energy is again harnessed to decrease the redox potential of the P700 redox pair to approximately -1.3 V.
- Electrons are finally transferred, down a redox potential gradient to NADP, to produce NADPH + H^+.

It is worth re-emphasising the significance of this scheme and the concept involved. Remember that the formation of C-C bonds is not net reduction. So all the reduced bonds in the glucose synthesised by photosynthesis are the result of a mechanism by which light energy is used to reduce $NADP^+$, using electrons from water. And these reduced bonds represent at least 75–80% of the reduced bonds in the fuels used by animal cells.

The only part of the reduction side of Figure I.1 that remains to be explained is the ATP that is used as a reactant in two reactions of the Calvin cycle (Figure 6.1.A).

6.3 The synthesis of the ATP required by the Calvin cycle

ATP is a reactant in reactions 2 and 15 of the Calvin cycle. The production of this ATP is not a conceptual or energetic problem, i.e. it could simply be provided by oxidising some of the glucose synthesised, through the normal oxidation pathways that are present in plant cells. The problem however, is one of quantity and compartmentalisation.

- There are a lot of chloroplasts in some plant tissues, and there is some argument as to whether the ratio of mitochondria to chloroplasts in these tissues could supply ATP at a sufficient rate to support the Calvin cycle in the chloroplasts.
- Even if the mitochondria could supply ATP at a high enough rate, reactions in the stroma of the chloroplasts do not have access to mito-

chondrial ATP. The Calvin cycle occurs in the stroma of the chloroplast (Figure 6.2.A), and there is no mechanism for transporting cytoplasmic ATP into the stroma of the chloroplast.

This combination of problems is solved by the synthesis of ATP in the stroma of the chloroplast, as a result of the electron transfers associated with $NADP^+$ reduction (Figure 6.3). The mechanisms involved are instructive as they are almost exact repetitions (in a plant cell and in a chloroplast) of two mechanisms that were originally encountered in the mitochondria of animal cells.

- The ATP is synthesised via a proton gradient, as described for the mitochondria in section 4.1. As in the mitochondria, the change in free energy as the electron transfer reactions between P680 and P700 proceed towards equilibrium, is conserved as a proton gradient across the thylakoid membrane (a high concentration in the lumen). Unlike in the mitochondria, the gradient is also produced as a result of proton production by the splitting of water in the lumen, and the uptake of protons by the reduction of $NADP^+$ in the stroma (Figure 6.2.B). Note that the system can also engage in what is called 'cyclic electron flow'. In this case $NADP^+$ is not reduced, the electrons are transferred from excited P700 to the cytochrome bf complex (Figure 6.3) which then transfers them to ground state P700, and the cycle repeats. This is how the high rates of ATP synthesis in chloroplasts (mentioned above) can be achieved.
- The proton gradient is then used by an ATP synthase to reverse the reaction of ATP hydrolysis. This is the structural repeat (repeating theme: structural repeats: ATP synthase) that was mentioned at the end of section 4.1. The ATP synthase in chloroplasts is remarkably similar in subunit composition, structure and function, to that in the mitochondria.

7 Reiteration, Variation and Application

7.1 Figure I.1 revisited

With the benefit of the hindsight provided by the previous six chapters, it is now possible to appreciate fully the complete oxidation/reduction cycle shown in Figure I.1.

The following refers to Figure I.1. Words taken from the figure are written as they are in the figure, and are in inverted commas.

- The 'Fuels (reduced)' comprise carbohydrates and lipids. The reduction is in the form of reduced bonds, which are C-C and C-H bonds.
- Quantitatively significant *'reduction'* only takes place in plants, and produces carbohydrate and oxygen. A degree of *'reduction'* can take place in animal cells, but the starting point is a compound (the acetyl group of acetyl CoA) that is already 70% reduced. In animal cells therefore, reduction is not a quantitatively significant process in terms of producing reduced bonds. It is more concerned with interconversion between types of fuels,[1] and with producing storage forms of the fuels. Because interconversion between fuel types is possible, the reduced bonds synthesised by plants are available to both plant and animal cells as either carbohydrate or lipid.
- The reduced bonds of the fuels are oxidised through a series of pathways that comprise preparation and oxidation reactions. The electrons extracted during the *'oxidation'* reactions are temporarily stored in specialised electron carriers and ultimately donated to 'Oxygen', which is produced by the water-splitting enzyme complex in chloroplasts. This oxidative scheme is the process which *'harnesses energy'* in the form of 'ATP'. The process can be considered as a reaction proceeding towards equilibrium. As it does so energy is conserved in the form of a disequilibrium between ATP and its hydrolysis products.
- The *'oxidised'* products are represented by compounds in which all electrons are being shared with oxygen. The carbon of the fuels appears as 'CO_2', the hydrogen and oxygen appear as 'H_2O'.
- The oxidation/reduction cycle represented by Figure I.1 is completed when these oxidised products are used in the process of photosynthesis

('*reduction*'). Photosynthesis is powered by light energy and uses electrons from water to reduce CO_2 to produce the 'Fuels (reduced)'.

7.2 Variations of the basic scheme

In Chapter 3, variations of the basic metabolic scheme were briefly mentioned (Box 3.1). In this section the issue will be explored in more detail. The message is that there are certainly many different forms that oxidative hardware can assume, but that no new concepts need to be invoked. The concepts involved have all been previously encountered and relate to the transfer of electrons along a redox potential gradient, from one redox pair to another, with the concomitant synthesis of ATP from ADP. This section will therefore reiterate some of the concepts that have been discussed previously, such as redox potential, innocuous end-products, and energy availability from a reaction out of equilibrium, to name a few.

7.2.1 PRODUCING END-PRODUCTS OTHER THAN CO_2 AND H_2O

The production of lactate is a good example of this variation of the basic theme, and is covered in detail in section 4.4. To briefly reiterate, one mole of glucose (twelve reduced bonds) produces two moles of lactate, and lactate comprises six reduced bonds. So if all the reduced bonds in glucose are accounted for in lactate, how does the conversion of a mole of glucose to two moles of lactate result in the production of two moles of ATP? This question was addressed in section 4.4, and the answer is that this pathway does in fact involve some oxidation, courtesy of the unique character of an aldehyde oxidation, and the fact that not all reduced bonds are created equal (see concept: energetics of an aldehyde oxidation). So the mechanism of ATP synthesis during lactate production does not deviate from the usual set of concepts, but is an interesting variation in that the product is not innocuous. This raises a further question: why would a cell opt for such an energy-producing pathway?

There are two situations in which lactate production is the best choice of pathways. The first has already been mentioned in section 4.4. Lactate production is a rapid way of producing ATP and suits intense forms of exercise in which rates of ATP demand cannot be matched by rates of oxygen delivery. Innocuous end-products are not therefore an option. ATP synthesis via lactate production is, however, limited by lactate accumulation and these forms of exercise are therefore necessarily of a short-term nature (e.g. a 100 m sprint). The accumulated lactate is subsequently completely oxidised (by conversion back to pyruvate, and thence into the Krebs cycle) at low rates that match those of oxygen delivery. The second situation in which lactate production is a viable option is that in which the lactate produced can be constantly and reliably exported to another tissue (or animal) for oxidation. An interesting example of this is the human brain, which converts about 10% of the glucose it consumes to lactate. There is still speculation about the rationale for this lactate production, but as the brain is always reliably perfused (thus ensuring a constant oxygen supply for glucose oxidation, and a reliable export system via the blood), the lactate is exported as fast as it is produced. So in this case, lactate can be considered an innocuous end-product, at least from the point of view of the brain. And one could envisage a parasite like a tapeworm, in the anoxic environment of the gut, doing the same thing and exporting the lactate to the host animal for complete oxidation.

In fact some parasitic worms do this, but their pathways do not necessarily stop at pyruvate or lactate. These pathways (which include that of lactate production) are defined by two characteristics. First, the oxidation of the substrates is only partial; and second, ATP production does not involve the mitochondria, but reactions such as decarboxylations in which ATP and ADP are co-substrates. One such pathway ends in acetate, which has only four reduced bonds and thus reflects more oxidation than a pathway that ends with pyruvate. And in keeping with the somewhat flexible principles of reduced bond/ATP stoichiometry, this pathway produces four moles of ATP per mole of glucose. This is only one of a number of pathways that have evolved in various animals, to enable them to maximise energy output from glucose in anoxic environments. In some animals, pyruvate is further metabolised for reasons that do not involve energy production. For instance, ethanol is produced in goldfish as an easily excreted end-product, and some bivalve molluscs produce alanopine, a process which essentially ameliorates the detrimental effects of arginine that is produced concomitantly with pyruvate.[2] These may seem strange and 'new' pathways as first glance, but, *nota bene*, there are no new concepts involved.

7.2.2 USING ELECTRON ACCEPTORS OTHER THAN OXYGEN

Oxygen is not a unique electron acceptor in energy metabolism. It happens to be an effective electron acceptor because it has such a high affinity for electrons, and because the product of the reaction in which oxygen accepts electrons is water. But in terms of lactate production, pyruvate is actually the ultimate electron acceptor. Look back at Figure 4.9. The aldehyde glyceraldehyde-3-phosphate is oxidised to 1,3 bisphosphoglycerate and the electrons are transferred to NAD, and from there to pyruvate to produce lactate. Electron acceptors can be any compound that is available, as long as the ΔG value for the transfer reaction is negative (Figure 2.4, Box 2.2). This is amply demonstrated by the organisms that live in the highly reduced layers of benthic silt. These bacteria and lower invertebrates use SO_4^{2-} as the electron acceptor in a pathway that probably looks something like the equation below (CH_2O = carbohydrate).

$$CH_2O + SO_4^{2-} + H^+ \rightarrow H_2S + HS^- + H_2O + CO_2\;^3$$

This strategy is not without its problems however, as the hydrogen sulphide product can be a potential problem; it is a potent inhibitor of one of the components of the electron transfer system (Complex IV, see Figure 4.1). This problem has been addressed in the literature, but is not discussed here.

7.2.3 USING NON-ORGANIC COMPOUNDS AS THE SOURCE OF ELECTRONS

The penultimate variation that we will discuss is that in which the reduced bonds themselves derive from sources that have not previously been considered. So far, our schemes have involved C-H and C-C bonds in compounds such as carbohydrates, lipids and amino acids. But these bonds are also found in other organic compounds like methane. Enormous reservoirs of methane are buried beneath the sea floor. The specific microbes and pathways involved are not well understood, but it is estimated that 5–20% of the total annual methane flux to the atmosphere is consumed by methane oxidation to produce energy.[4]

Methane is really just a simple form of a carbohydrate or a lipid (although highly reduced), but reduced bonds are not restricted to the C-C and C-H form, neither are they restricted to organic molecules. As long as a donator and an acceptor of electrons are available, with suitable redox potentials, it is possible for a biochemical pathway to extract the energy by transferring the electrons. Sulphur has already been implicated (as SO_4^{2-}) as an electron acceptor, but is also involved in energy production as an electron donor. H_2S can be found in some regions of the sea floor in large amounts and concentrations. Sulphide is a potent reductant with redox potentials between -0.4 and -0.1 V, depending upon the redox partner involved. Again the mechanisms and reaction involved are still unclear, but a variety of non-bacterial organisms, including some of the inhabitants of hydrothermal vents, are believed to use H_2S as an energy source.[5]

Finally, some bacteria use novel electron donors *and* acceptors. These bacteria are found in anoxic marine sediments and water which display high concentrations of NH_4^+. They use NH_4^+ as the electron donor and NO_2^- as the acceptor and catalyse the following reaction.

$$NH_4^+ + NO_2^- \rightarrow N_2 + 2H_2O$$

In some areas it is calculated that this reaction could account for up to 35% of the total N_2 formation in the water column.[6]

7: REITERATION, VARIATION AND APPLICATION

7.3 Application of the concepts: weight control in humans

The two most important points about weight control are simple. They relate to now familiar concepts and are listed below.

1. A person can only put on weight by taking in reduced bonds through the mouth.
2. A person can only decrease in weight by metabolising (oxidising) the store of reduced bonds that comprise the weight.

If we translate reduced bonds into kilojoules (kJ) (the familiar unit of energy found on food packets), these two points lead to the next three points which are:

3. To gain weight, a person must take in more kJ than they metabolise.
4. To lose weight, a person must take in less kJ than they metabolise.
5. To maintain an even weight, a person must take in the same number of kJ as they metabolise.

So a day for the average person (Mr Fit N Healthy) would look something like this. Mr Healthy at rest will be using energy at about the same rate as a 100 W light bulb (Chapter 1, section 1.5), which is approximately 8000 kJ per day. As we have discovered, lipid contains almost twice as much energy per gram as carbohydrate (Figure 4.10). So to replace the reduced bonds that are oxidised to produce the 8000 kJ, Mr Healthy must take in about 200 g of lipid, or 360 g of carbohydrate, or some mixture of the two. In this case his weight will remain constant (see point 5 above). If Mr Healthy does some exercise during the day, his energy utilisation will increase. For example, an aerobic session will cost about 2500 kJ. If this extra oxidation is not accounted for by increased intake, Mr Healthy will lose about 60 g, which represents the amount of lipid required to produce 2500 kJ (point 4 above).

So the rules governing weight control are simple, as outlined in points 3–5 above. It should therefore be easy to maintain, or either lose or gain weight. But this is obviously not the case, as obesity is a problem, and an increasing one in many countries. Since obesity has such a profound detrimental effect on health and lifestyle, it behoves us to ask why it is so difficult to lose weight.

In many cases the problem is probably willpower. These people understand points 1–5 (above), but do not have the willpower to both exercise regularly and to curtail their intake. But for many people it is a misunderstanding of points 1–5 (above), a problem that can be exacerbated by questionable advertising, or advice that is unnecessarily complex. This leads to confusion about what should be done in terms of exercise and diet. What follows makes use of some of the concepts outlined previously, and comprises some useful facts, a clarification of some common misunderstandings, and some advice, all of which should simplify, demystify and facilitate the process of weight control.

- All kJ taken in are used to either fuel the metabolic rate or are eventually stored as lipid (fat), as the capacity of humans to store carbohydrate is very limited.[7] A kJ is a kJ, regardless of whether it is contained in lipid, carbohydrate or protein. So if more kJ are taken in than are used, regardless of the form this surplus takes, it will end up as stored lipid.

- It is simpler if nutrition and weight control are not tackled together. It is probably a healthier alternative to take in kJ as a banana rather than as a bag of chips. But in terms of weight control, 100 kJ of banana converts to the same amount of lipid as 100 kJ of chips. Qualitative changes to the diet can be very difficult, so, in terms of weight control, changes are best kept simple. So if a person's diet consists of chips and hamburgers, the simplest way for this person to lose weight is to cut down on the quantity of chips and hamburgers. The nutrition issue can be dealt with later.

- Claims such as 'fat-burning aerobics' may contain an element of truth, but are not relevant to the issue of weight loss. It is true that some forms of exercise use more lipid than carbohydrate, and vice versa, but the only pertinent number is how many kJ are used, and this is determined only by the intensity and duration of the exercise. Regardless of which sort of fuel is used during a particular exercise regime, the kJ will ultimately come from the lipid stores. Depleted carbohydrate stores are always replaced from the diet, as they are obligate fuels for sensitive tissues like the brain.

- There are three ways that Mr Healthy can lose weight. He could raise his resting metabolic rate by 20%, from 8000 kJ to 9600 kJ per day. If he was doing no exercise, and his intake remained at 8000 kJ per day, the increase in the resting metabolic rate would result in a weight loss of 102 g per day. But raising the resting metabolic rate is difficult. It can be done to some extent by exercise, as the metabolic rate tends to remain high for a period after exercise. But even after an eighty-minute session of extremely intense exercise, the resting metabolic rate is only about 10% higher than normal after six hours, and is back to normal after nine hours. A forty-minute exercise session of the same intensity has no effect on resting metabolic rate. Alternatively, Mr Healthy could do two average-intensity one-hour aerobic sessions a day and use the exercise to lose weight; in this case he would lose about 125 g per day. But fat is an alarmingly efficient fuel. Tiny amounts of it power the migration of birds and insects over vast distances, and it keeps hunger strikers alive for months. Consequently, an adult human needs to do about sixteen hours of very high-intensity aerobics to burn 1 kg of fat. So the answer is inescapable: *the easiest way to lose weight is to eat less*. If Mr Healthy cut his food intake by half, he would lose 162 g per day. But 4000 kJ is equivalent to about thirteen average slices of bread. This is a major quantitative change in the diet. The most effective way to cut down on food intake is to decrease it by a very small amount over the very long term, i.e. a very small, but permanent change. Mr Healthy would be best to target an item like breakfast cereal, which is eaten in the same quantity each day. He would weigh a normal day's serving, and from then on cut down the serving by only 10–15%, i.e. instead of having 70 g, have 60 g. If every other aspect of the diet and exercise regime remain unchanged, and assuming that the person's weight was steady at the beginning, the weight will decrease with time, slowly but inexorably. And the change in the diet will be virtually undetectable.

- Finally, there are no mysterious or 'alternative' ways of losing weight. The only process that depletes lipid stores is fuel oxidation, which leads back to a concept, and a timely note on which to end the book. Remember that lipid stores represent carbon, hydrogen and oxygen, and that these elements can only exit the body as carbon dioxide and water (the products of lipid oxidation). So to lose weight, one must excrete the two molecules that represent the innocuous end-products (Box 3.2; concept: innocuous end-products) which are key to the operation of the fuel oxidation system that sustains cells in the long term in an aerobic environment.

Notes

[1] Both animal and plant cells can convert the reduced bonds of glucose into the lipid form. Glucose is first converted to pyruvate (Figure 4.9), which is then converted to acetyl CoA (Figure 4.8), which is the starting point for fatty acid synthesis (Figure 5.1). Most animal cells cannot convert lipid into carbohydrate, but plants can. This process, the glyoxylate cycle, is described in biochemical texts.

[2] This subject is covered in more detail in Hochachka, P W, and Somero, G N, *Biochemical Adaptation: Mechanism and Process in Physiological Evolution*, New York, Oxford University Press, 2002.

[3] See Hochachka and Somero, 2002.

[4] Orphan, V J, House, C H, Hinrichs, K, McKegan, K D, DeLong, E F, 'Multiple archaeal groups mediate methane oxidation in anoxic cold seep sediments', *Proceedings of the National Academy of Sciences*, 2002, 99, p.7663.

[5] Zierenberg, R A, Adams, M W W, Arp, A J, 'Life in extreme environments: Hydrothermal vents', *Proc. Nat. Acad. Sci.*, 2000, 97, p.12961; Grieshaber, M K, Volkel, S, 'Animal adaptations for tolerance and exploitation of poisonous sulfide', *Annual Review of Physiology*, 1998, 60, p.33.

[6] Dalsgaard, T, Canfield, D E, Petersen, J, Thamdrup, B, Acuna-Gonzalez, J, 'N_2 production by the anammox reaction in the anoxic water column of Golfo Dulce, Costa Rica', *Nature*, 2003, 422, p.606.

[7] The average human would have stores consisting of about 450 g of carbohydrate and 9000 g of lipid.

Figures

FIGURES

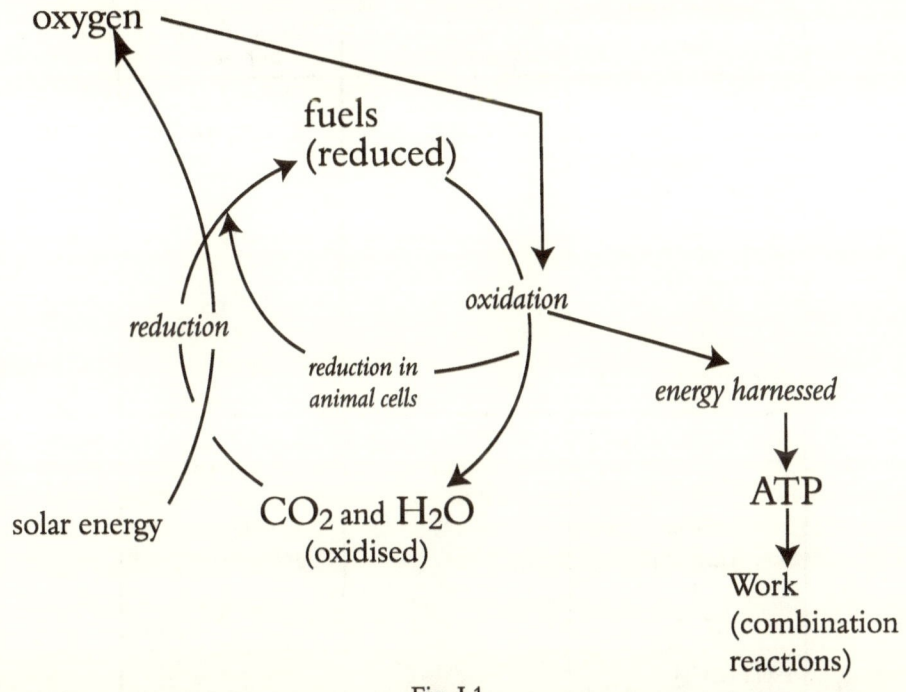

Fig. I.1

Figure I.1 The oxidation/reduction cycle in cells

- Extraction: fuels are oxidised and the energy is harnessed as ATP (adenosine triphosphate).[1]
- Utilisation: the ATP enables work[2] to be done.
- Incorporation: the products of fuel oxidation (CO_2 and H_2O) can be reduced, using energy from the sun, to produce the original fuels.
- Note that all of the reactions shown, only proceed (always towards the equilibrium position) if the reactants represent more total free energy than the products (see 1.1 point 3).

---

[1] ATP is a molecule involved in energy transfer between reactions. The characteristics of ATP (including the thermodynamics of its hydrolysis; see footnote 2 below) that render it suitable for its role in cellular energy transfer will be discussed in detail in later sections.

[2] Work in the cell that is powered by ATP, is done by incorporating ATP hydrolysis into a combination reaction that includes a reaction which would not otherwise proceed in the cell. The concepts involved in these reactions are discussed in detail in later sections.

Fuel	Reduced bonds per carbon	Reduced bonds per gram
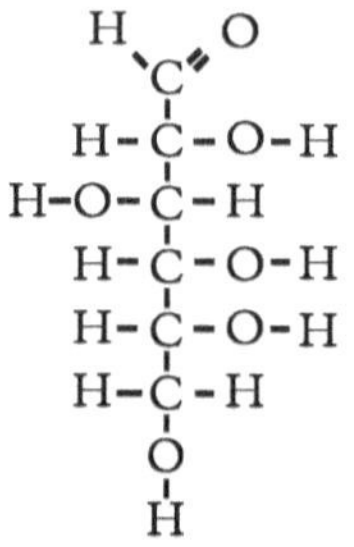 Carbohydrate (CH_2O)	2.0	4.0×10^{22}
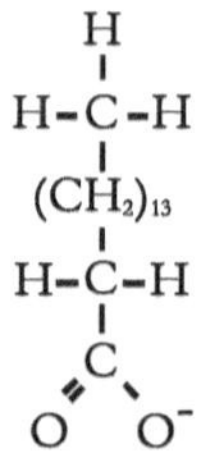 Lipid (CH_2)	2.9	1.1×10^{23}
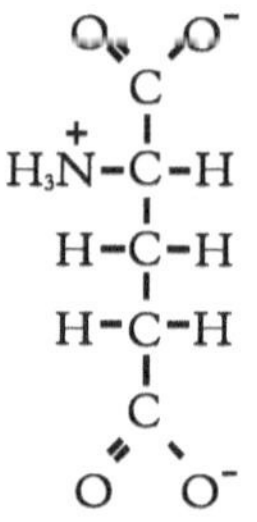 Amino acid (CH_2ON)	2.0	4.1×10^{22}

Fig. 1.1

Figure 1.1 The types of fuel molecules and their characteristics in terms of reduced bonds

- For the purposes of this book, C-C and C-H bonds are termed 'reduced bonds'.
- The biologically available energy in these molecules (glucose, palmitate and glutamate) is in the reduced bonds.
- The formulae in parentheses are the general chemical formulae for each class of molecule.

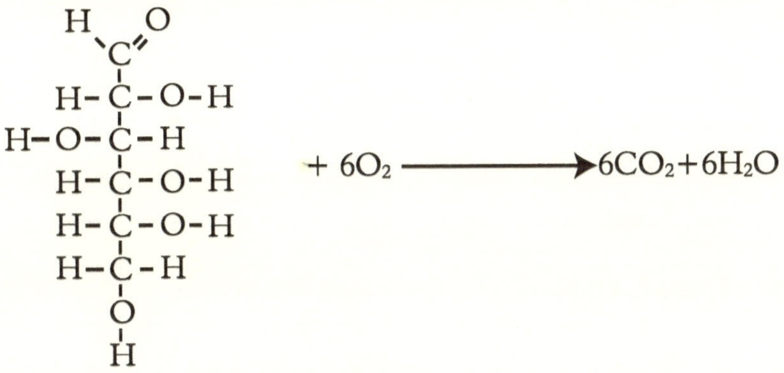

$$\text{glucose} + 6O_2 \longrightarrow 6CO_2 + 6H_2O$$

Fig. 1.2

Figure 1.2 The oxidation of glucose in terms of oxidation of reduced bonds

- All the reduced bonds (C-C and C-H) in the glucose molecule have been oxidised in this reaction.
- In the products, all bonds comprise electrons being shared with oxygen.

FIGURES

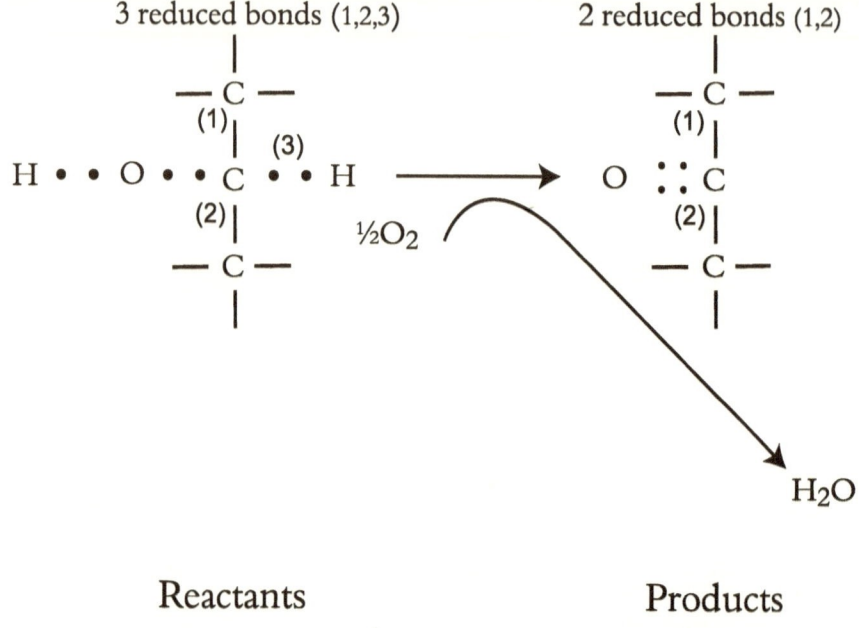

Fig. 1.3

Figure 1.3   The oxidation of a reduced bond

- For the purposes of this example consider only the reduced bonds in the vicinity of the carbon involved in the oxidation, i.e. bonds 1, 2 and 3. The rest of the fuel molecule is not altered in the reaction.
- The total number of reduced bonds always decreases by one as these oxidation reactions proceed.
- After oxidation of bond 3, the electrons that comprised reduced bond 3 are shared with oxygen, either in the ketone group, in the partially oxidised fuel molecule, or in water (see products).

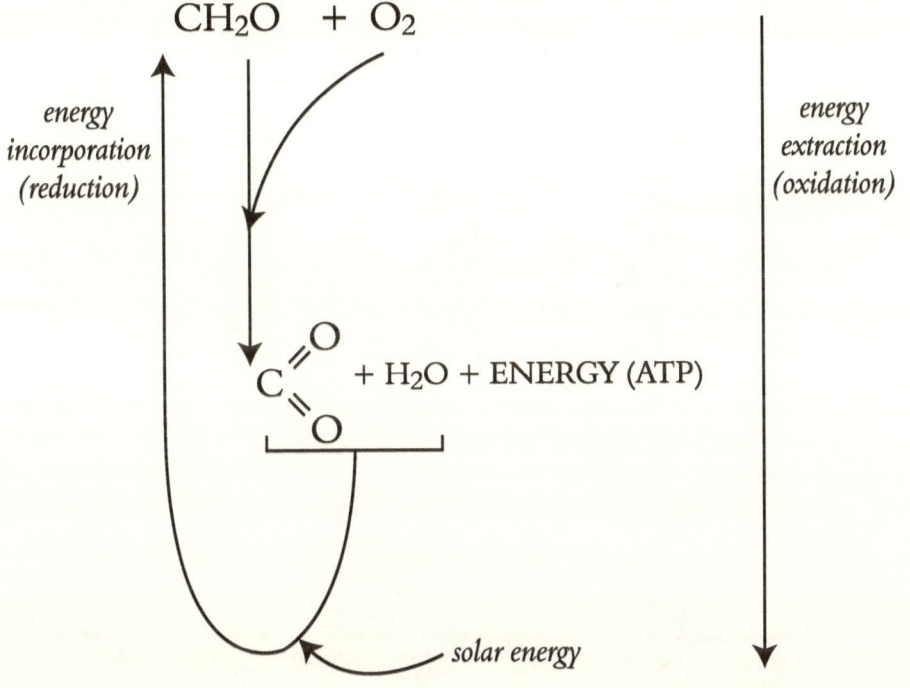

Fig. 2.1

Figure 2.1 Energy extraction and incorporation viewed as a reversible oxidation/reduction reaction

- CH_2O represents a carbohydrate (see Figure 1.1).
- Energy is extracted by oxidation. The C-C and C-H bonds are oxidised to produce CO_2 and H_2O and energy. This reaction occurs in the cell, but also occurs in a wood fire. In the case of the fire, ATP production is bypassed and the energy output manifests immediately as heat.
- In the incorporation phase, powered by sunlight, the CO_2 is reduced using electrons from the H_2O.

1. Alcohol to a ketone (Krebs cycle)

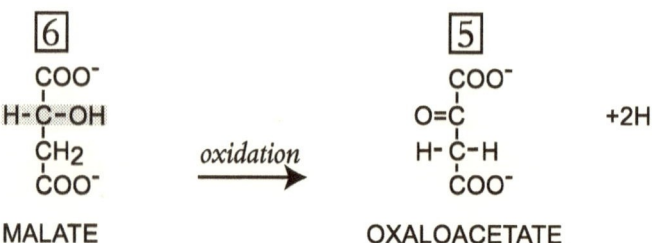

2. Alkane to an alkene (Krebs cycle)

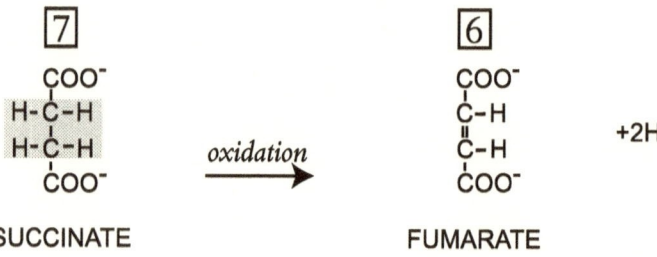

3. Aldehyde to an acid (glycolysis)

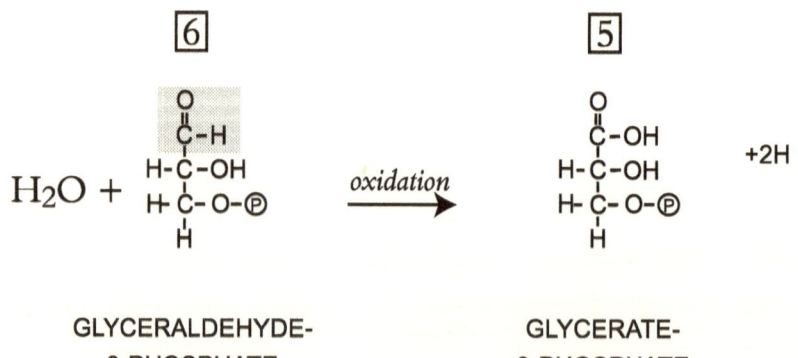

Fig. 2.2

Figure 2.2 Examples of the three types of oxidation that occur in pathways of energy extraction

- The numbers in the squares above each compound represent the number of reduced bonds in the molecule. Note that the number decreases by one with each oxidation.
- The shaded portions represent the parts of the molecules which participate in the reaction, and which are oxidised as a consequence.
- The names of the molecules will take on more meaning when they are encountered in the detailed examination of the various metabolic pathways.

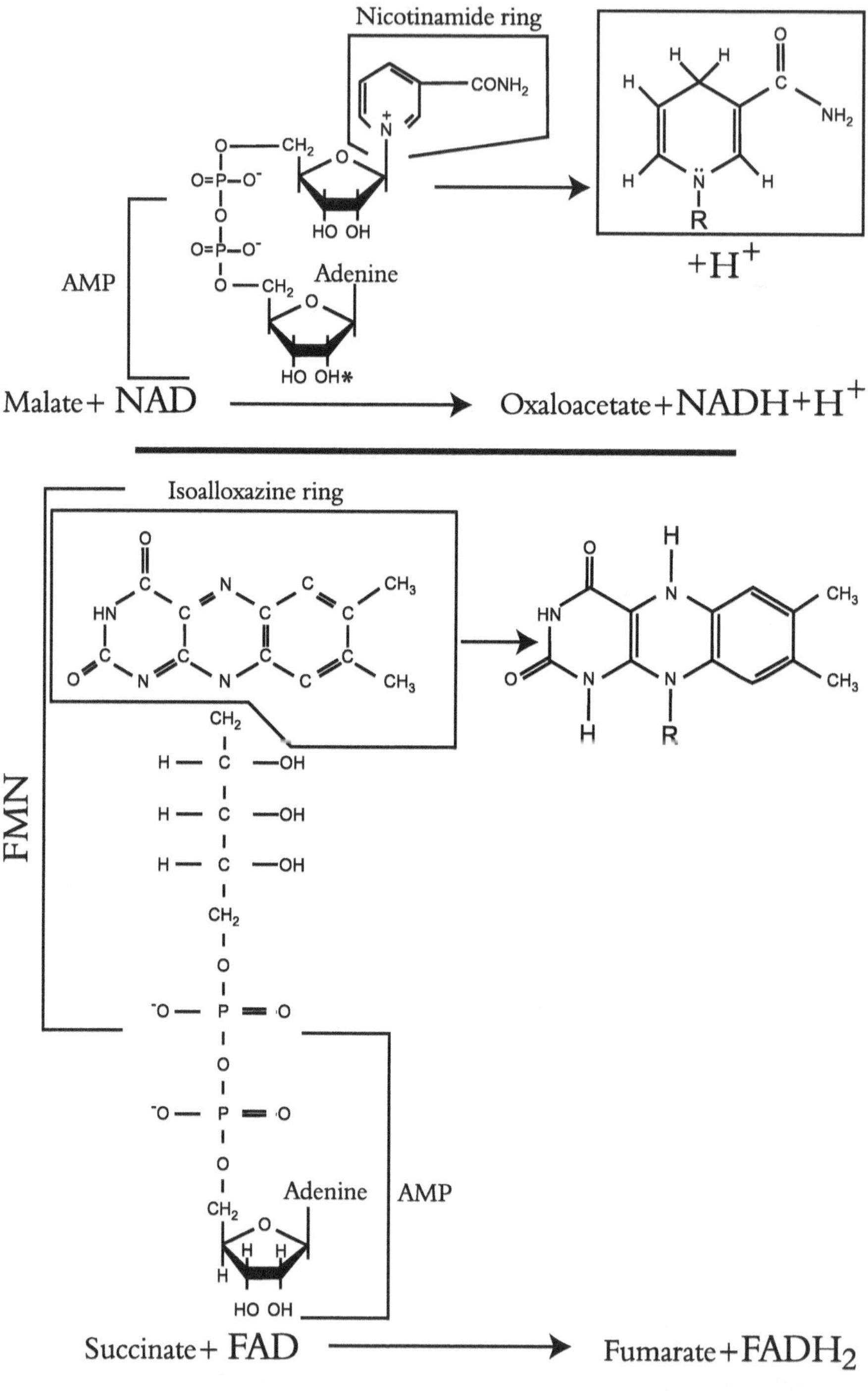
Nicotinamide ring
CONH2
O=P–O⁻
HO OH
Adenine
AMP
HO OH*
+H+
R
Malate+ NAD
Oxaloacetate+NADH+H+
Isoalloxazine ring
CH3
HN
FMN
Adenine
AMP
HO OH
Succinate+ FAD
Fumarate+FADH2

Fig. 2.3

Figure 2.3 NAD and FAD, electron carriers

- Two reactions are shown in this figure, one involving NAD (top) and one involving FAD. Only the structures of the electron carriers are shown.
- The part of the compound that carries the electrons is boxed.
- The structures on the right-hand side of each equation depict only that part of the electron carrier that carries the electrons; the rest of the molecule is designated by R.
- The names of the electron carriers describe the various parts of the molecules. NAD stands for nicotinamide adenine dinucleotide. There is a nicotinamide ring, an adenine base, and two nucleotides (base-sugar-phosphate). FAD is flavin (derived from riboflavin which is the isoalloxazine ring attached to a ribose) adenine dinucleotide. FMN stands for flavin mononucleotide and this molecule will be encountered as an electron acceptor in the electron transfer system in Chapter 4.
- The two electron carriers are shown performing their function as part of one of the oxidation reactions shown in Figure 2.2. Although it is not important for understanding the principles of how these carriers work, note for clarity that NAD^+ carries two electrons and one proton. This is why the reduced form is written as $NADH + H^+$. FAD carries two electrons and two protons.
- There is a repeating theme here (NAD and FAD specificity). With one exception, when NAD is involved, the oxidation is always an alcohol to a ketone or an aldehyde to an acid. When FAD is involved, the oxidation is always an alkane to an alkene.
- Adenosine monophosphate (AMP) is part of both molecules and is discussed in the text.
- In NADP (Chapter 5), the hydroxyl group marked with an asterisk is esterified with phosphate.
- Adapted from Matthews, C K, van Holde, K E, Ahern, K G, *Biochemistry*, third edition, San Francisco, Benjamin/Cummings, 2000.

oxygen/water	0.82
fumarate/succinate	0.031
H^+/H_2 (pH 0)	0
oxaloacetate/malate	-0.17
pyruvate/lactate	-0.19
$FAD/FADH_2$	-0.22
$NAD^+/NADH$	-0.32
$NADP^+/NADPH$	-0.32

Fig. 2.4

Figure 2.4 Standard redox potentials of some biologically important redox pairs (V, 25 °C, pH 7.0)

- The E'_0 values are an indication of the affinity for electrons of the oxidised form of the pair.

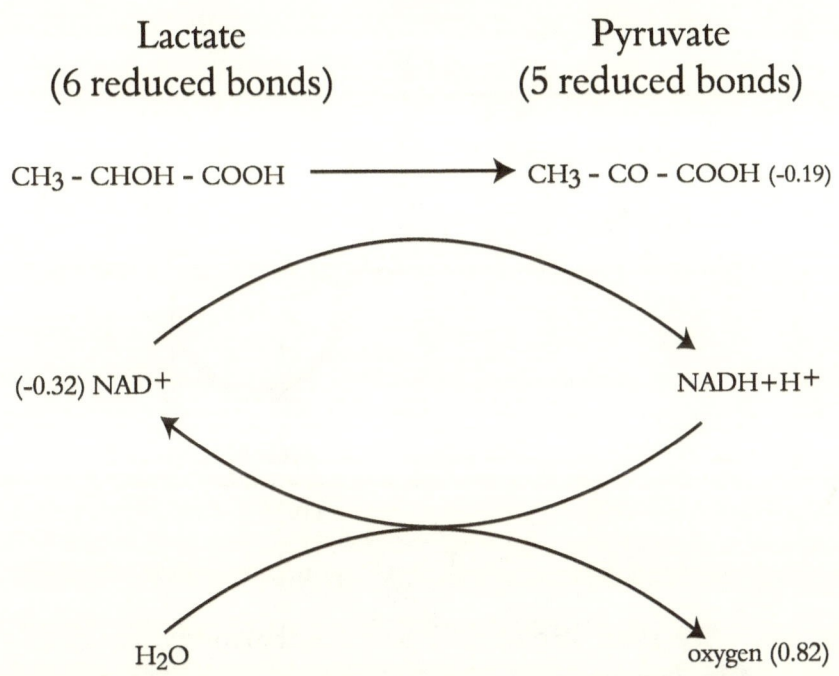

Fig. 2.5

Figure 2.5 The role of redox potential in electron transfer reactions

- The affinity for electrons (E'_0 in volts) is given in brackets next to the oxidised form.
- The transfer of electrons from the CHOH group of lactate to oxygen is possible despite the standard redox potentials which suggest otherwise (see Figure 2.4 and Box 2.2).
- Lactate is a typical fuel molecule, in essence, half a glucose.
- It contains C-C and C-H bonds and a CHOH group.
- Note the numbers of reduced bonds in lactate (reduced) and pyruvate (oxidised).
- Note that NAD^+ is involved in the oxidation of the alcohol group.

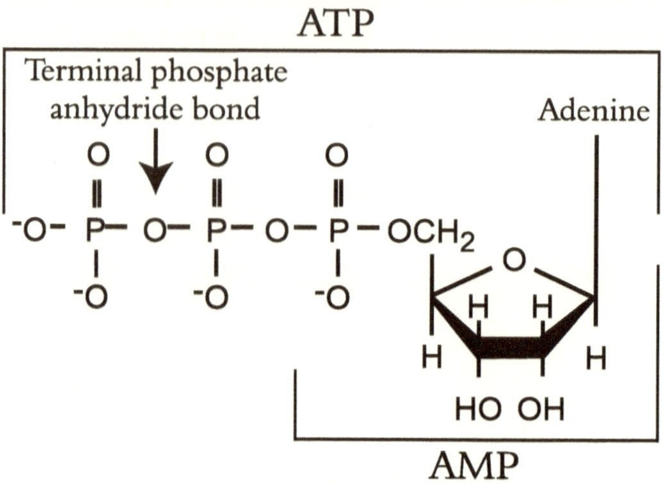

ATP

[Structure diagram: Terminal phosphate anhydride bond indicated on leftmost phosphate; three phosphates linked $-O-P(=O)(-O^-)-O-P(=O)(-O^-)-O-P(=O)(-O^-)-OCH_2-$ connected to ribose sugar with HO OH groups, and Adenine attached. AMP portion bracketed on the right.]

Fig. 2.6

Figure 2.6 The structure of ATP and the thermodynamics of its hydrolysis

- ATP can be hydrolysed as per the following reaction:

$$ATP + H_2O \rightarrow ADP + Pi + H^+$$

- $K_{eq} = 2 \times 10^5$.
- $\Delta G^{o\prime} = -30.5$ kJ/mole.
- $\Delta G = -51.8$ kJ/mole (will vary with cell and situation).
- The value of the [products]/[reactants] ratio in the cell (termed the mass action ratio) varies with the cell and the state of the tissue, but is roughly 2×10^{-4}.
- The reaction is therefore a factor of approximately 10^9 out of equilibrium in the cell.
- Adapted from Lehninger, et al., *Principles of Biochemistry*, New York, Worth Publishers, 1993.

1. glucose + Pi $\longrightarrow$ glucose-6-phosphate + H_2O
$K_{eq} = 3.9 \times 10^{-3}$
$\Delta G^{o'} = 13.8$ kJ/mol

2. glucose + ATP $\longrightarrow$ glucose-6-phosphate + ADP + H^+
$\Delta G^{o'} = -16.7$ kJ/mol

3. ATP + H_2O $\longrightarrow$ ADP + Pi + H^+
$\Delta G^{o'} = -30.5$ kJ/mol

Fig. 2.7

Figure 2.7 The role of ATP hydrolysis in enabling otherwise biologically impossible reactions

- Glucose is phosphorylated in the cell to produce glucose-6-phosphate. This is the first step in glycolysis and is crucial for glucose entry into all the pathways of glucose metabolism. But because the K_{eq} for reaction 1 is so low, the reactants would have be at impossibly (in a cell) high concentrations in order for this reaction to proceed as written, i.e. for ΔG to be negative.
- Instead, reaction 2, catalysed by the enzyme hexokinase, is used. The enzyme accepts both glucose and ATP as substrates and therefore uses the large negative $\Delta G^{o'}$ of the ATP hydrolysis reaction (3) to overcome the positive $\Delta G^{o'}$ of reaction 1. Reaction 2 occurs at cellular concentrations of approximately 3 mM glucose, 1.7 mM Pi and 0.2 mM glucose-6-phosphate.
- Another way of explaining why the combination reaction 2 has such a large negative $\Delta G^{o'}$, is that the $\Delta_f G^o$ of the phosphate anhydride bond of ATP (which is hydrolysed during the reaction) is much higher than that of the phosphate ester bond that is synthesised in glucose-6-phosphate. The 'extra energy' appears in the value of the K_{eq}, which ensures that this reaction will occur whenever it is required, not only when reactant and product concentrations are appropriate. This concept is termed the coupling coefficient and is discussed by Atkinson (1977; Chapter 1, note 3).

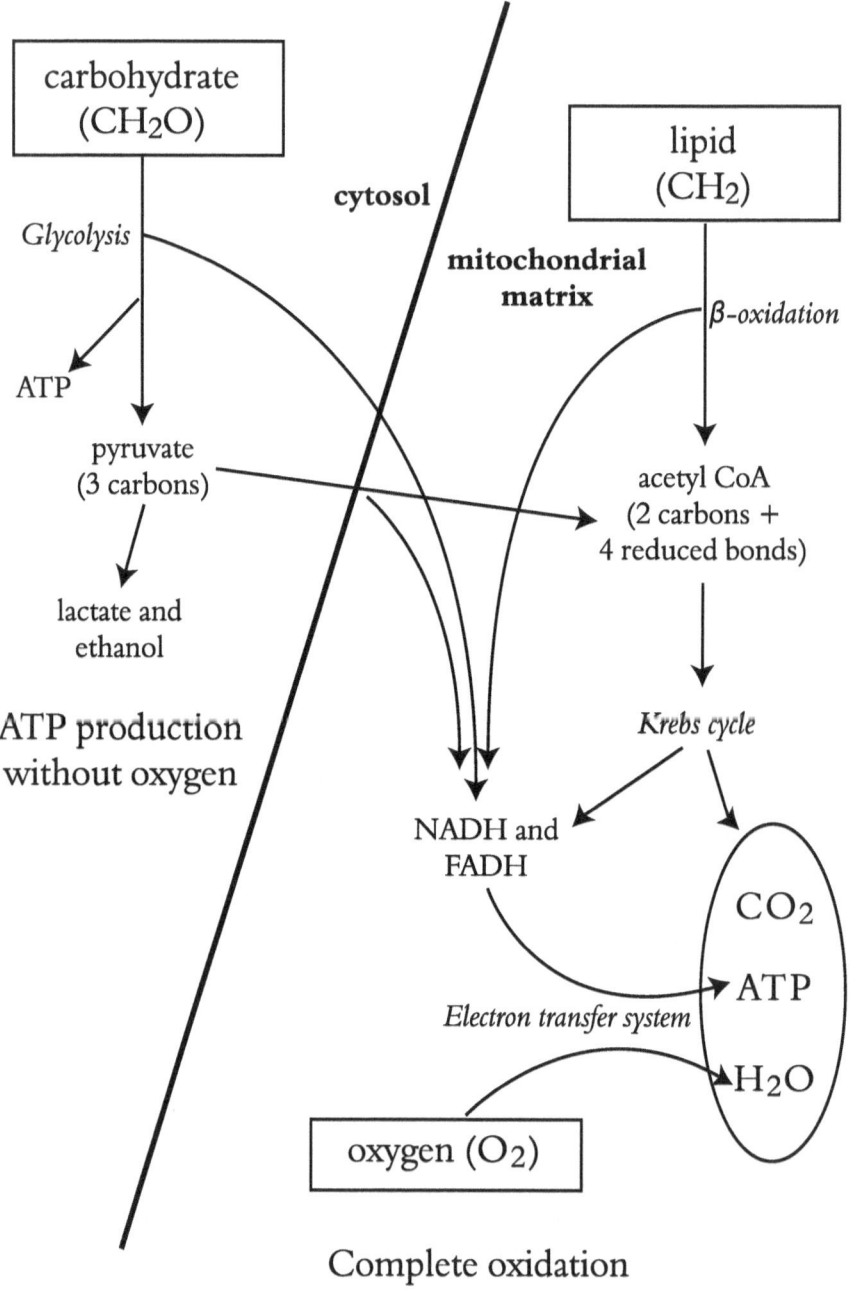

Fig. 3.1

Figure 3.1   An overview of the pathways involved in the oxidation of carbohydrate and lipid fuels

- The reactants for the complete oxidative process are in boxes; the products are in a circle.
- The names of the pathways are in italics; there are four of them.
- Note that complete carbohydrate oxidation can be summarised, using the boxes and circles in this figure, as:

$$CH_2O + O_2 \quad CO_2 + H_2O + ATP^3$$

- Carbohydrate and lipid oxidation become a common pathway at acetyl CoA which is produced in the mitochondrial matrix.
- The majority of the reduced electron carriers are produced through oxidation in the Krebs cycle.
- The electron transport system produces the major part of the cell's ATP and uses the oxygen in the equation above.

---

[3] As shown in Figure 2.1.

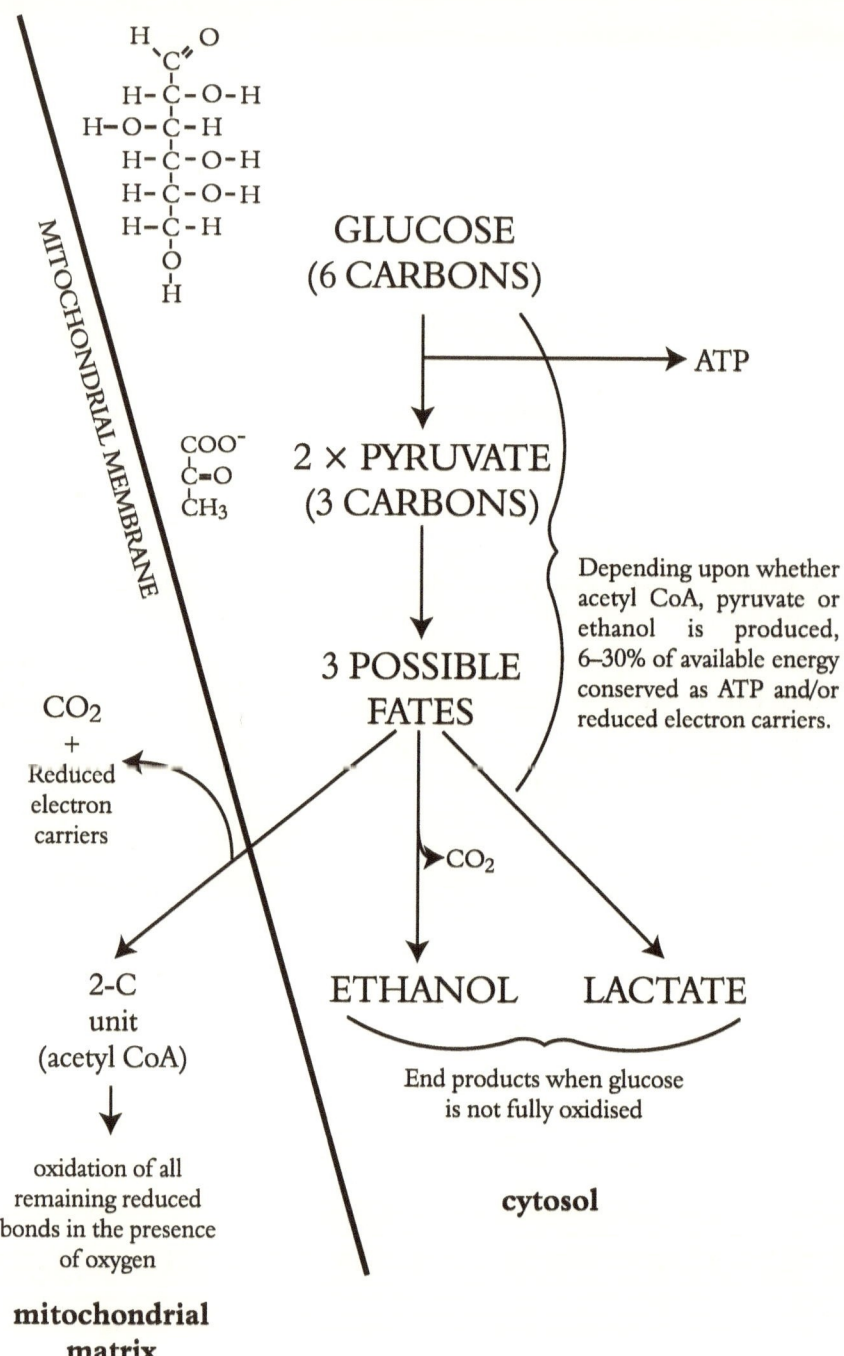

Fig. 3.2

Figure 3.2   A summary of glycolysis

- Glycolysis comprises the initial preparation process for glucose oxidation. In this case the pyruvate is converted to acetyl CoA, which is then further oxidised by the Krebs cycle.
- Glycolysis is also a mechanism by which ATP can be produced in the absence of oxygen. When glycolysis is functioning in this role, the pyruvate can be converted to lactate or ethanol for example, and the result is ATP production without the involvement of oxygen. As stated in Box 3.1, there are other options for the fate of pyruvate in the absence of oxygen.

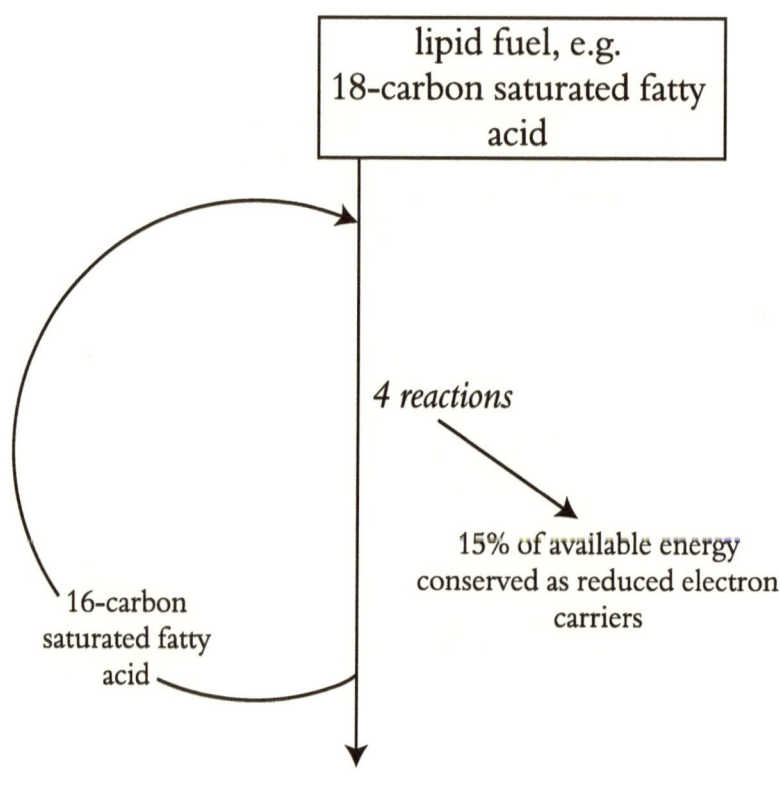

Fig. 3.3

Figure 3.3 A summary of β-oxidation

- Lipid fuels are processed through a set of four reactions, two carbons at a time.
- The two carbons exit as the acetyl group of acetyl CoA which enters the common (from glycolysis and β-oxidation) pool of acetyl CoA for the Krebs cycle.
- No ATP is produced, no oxygen is involved, and approximately 15% of the energy available in acetyl CoA is extracted and conserved as reduced electron carriers.

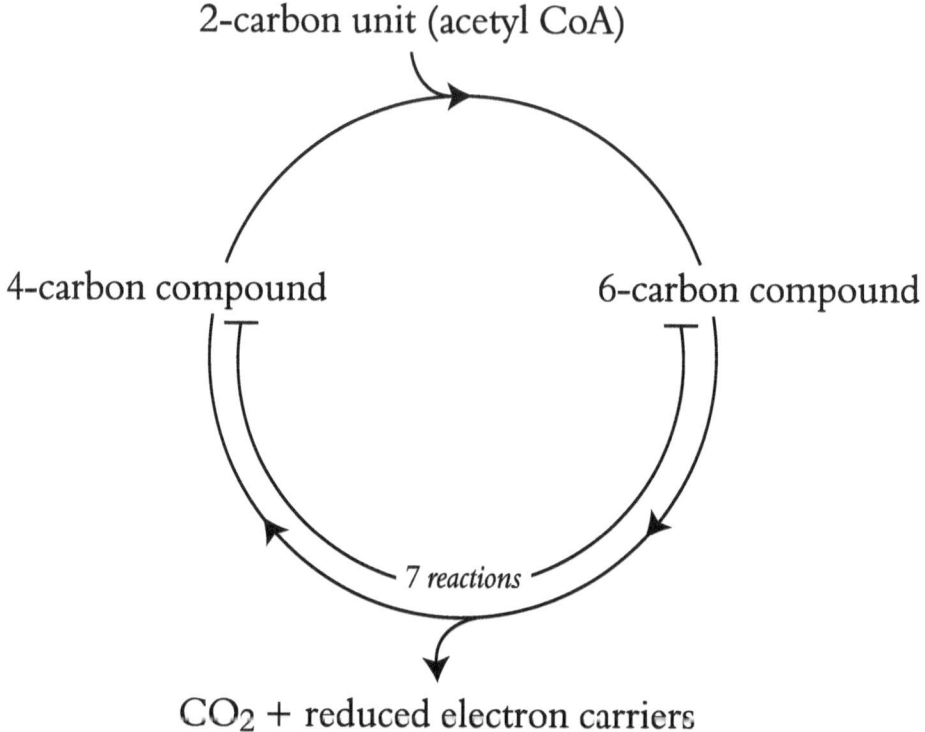

Fig. 3.4

Figure 3.4 A summary of the Krebs cycle

- The entry compound, carrying reduced bonds from carbohydrate and lipid fuels, is the 2-carbon acetyl CoA.
- The acetyl group from acetyl CoA is condensed with a 4-carbon compound.
- The resulting 6-carbon compound is the substrate for a series of reactions which oxidise the number of reduced bonds (four) equivalent to that which entered as the acetyl group.
- Carbon and oxygen (a stoichiometric equivalence to that entering as the acetyl group of acetyl CoA) are released as an innocuous end-product CO_2 (see Box 3.2) of oxidative metabolism.
- The energy from the oxidation of the reduced bonds is stored as reduced electron carriers.

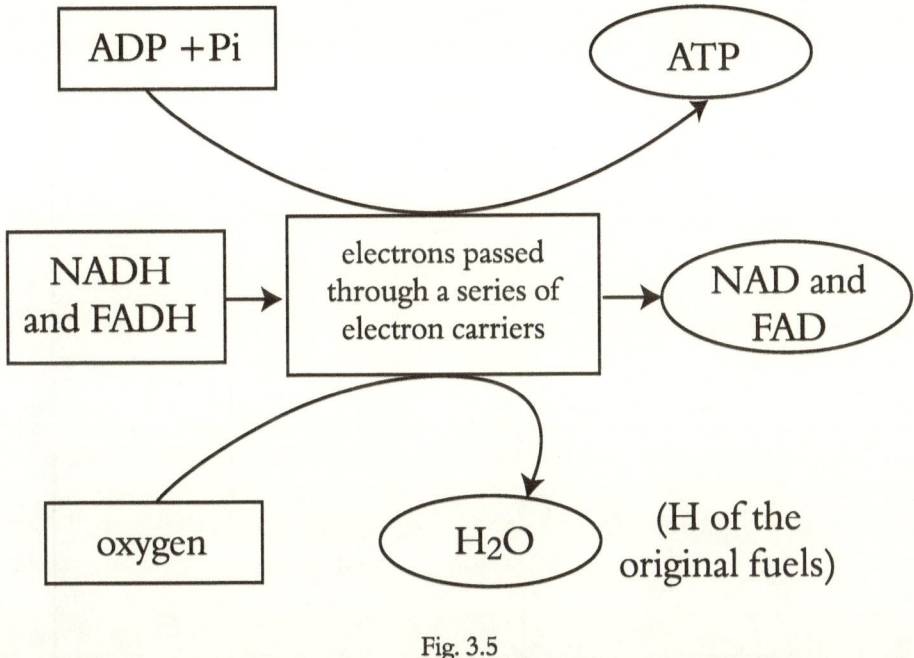

Fig. 3.5

Figure 3.5 A summary of the electron transfer system

- Reactants are in boxes, the products are in circles.
- The reduced electron carriers, carrying electrons from the reduced bonds of the fuels, donate the electrons to an electron carrier with a higher redox potential.
- The electrons are passed through a series of electron carriers with incrementally increasing redox potentials, eventually to be passed to oxygen to form water.
- In the process, by a mechanism that will be explained in Chapter 4, ADP is phosphorylated to ATP. Note that this reaction has a very large and positive ΔG (Figure 2.6), so cannot theoretically occur in the cell.
- The hydrogen from the original fuels appears as one of the innocuous products of oxidative metabolism, H_2O (see Box 3.2).

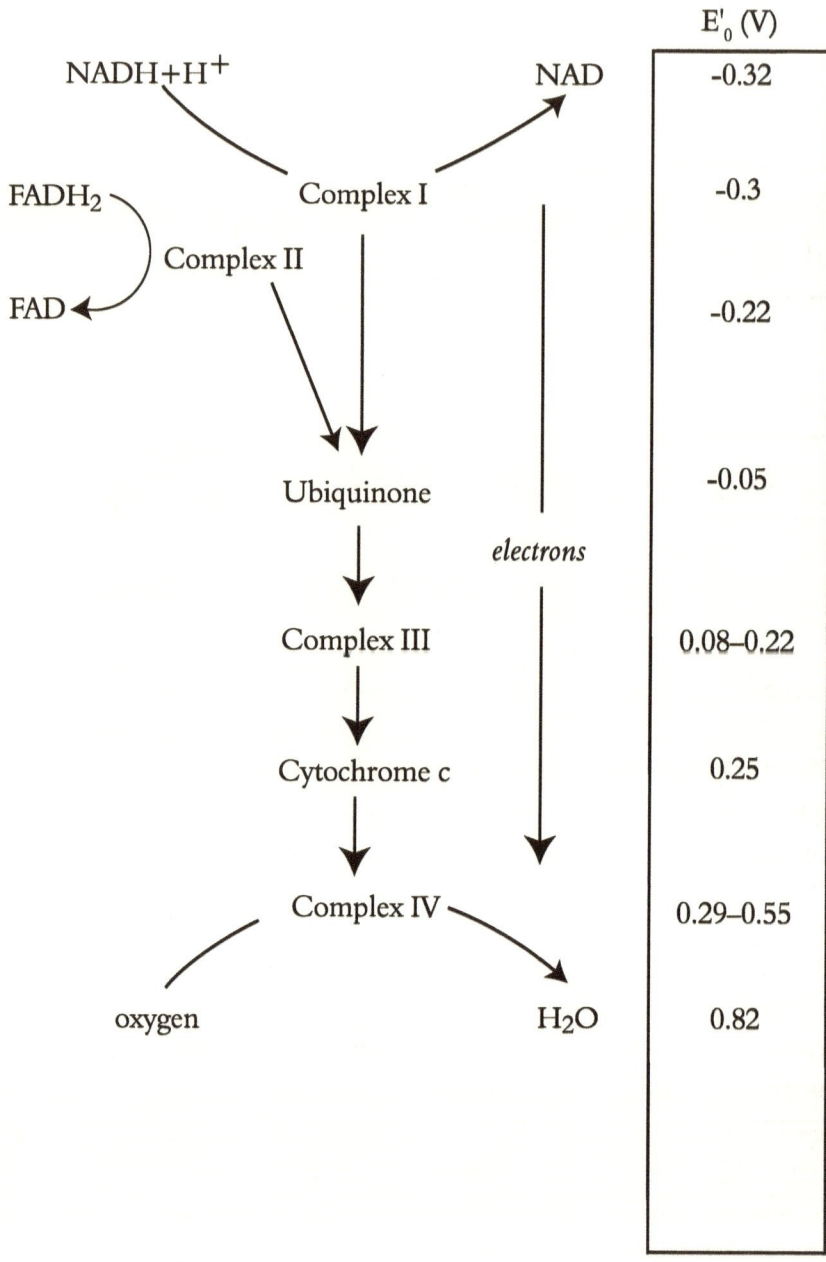

Fig. 4.1

Figure 4.1 A more detailed electron transfer system

- Electrons are transferred from NADH + H$^+$ and FADH$_2$, through the various electron acceptors/donators, eventually to oxygen.
- The complexes comprise large proteins (MW = 140–850 kDa) containing non-protein groups (prosthetic groups).
- It is the prosthetic groups in the complexes that have the relevant redox potentials, and which accept and donate electrons. These prosthetic groups include FMN (see Figure 2.3), iron-sulphur groups and copper ions. Consult a textbook if you wish to see exactly what reactions occur at each of these complexes.
- Ubiquinone has no protein component; it is a benzoquinone which can accept and donate electrons. Cytochrome c is a protein that contains a haem (porphyrin ring plus iron) group which is the electron acceptor and donator.
- The redox potentials on the right-hand side refer to the reaction in the scheme on the same level. For example the redox potential of the NAD/NADH + H$^+$ pair is -0.32V. NADH + H$^+$ can therefore donate electrons to the FMN of Complex I as the redox potential of the FMN\FMNH$_2$ pair in Complex I is -0.3V (see Figure 2.4 and section 2.2).

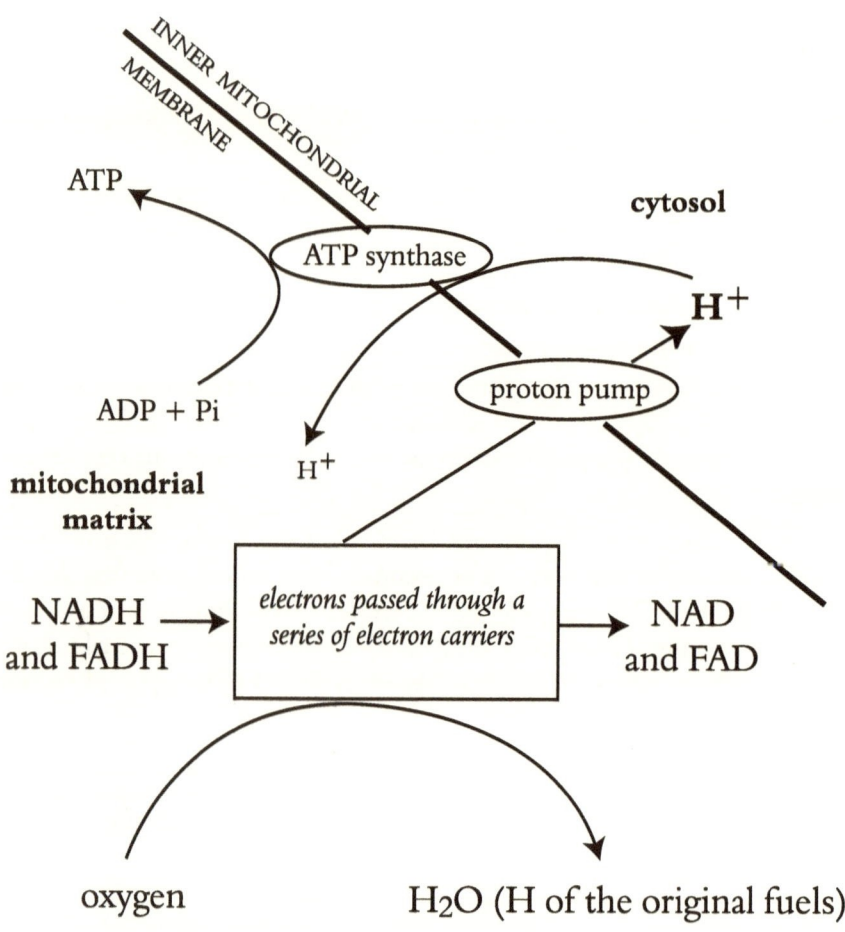

Fig. 4.2

Figure 4.2 ATP synthesis by the electron transfer system

- Start at NADH and FADH.
- Electrons are passed through a series of electron carriers, eventually to oxygen to form H_2O. The energy released from the reaction proceeding towards equilibrium is used to pump protons from the mitochondrial matrix to the cytosol, thus creating a proton gradient (represented by the large H^+ in the cytosol and the small H^+ in the matrix).
- The inner mitochondrial membrane is impermeable to protons thus forcing the protons to flow back into the matrix through a proton pore which is linked to an enzyme (ATP synthase) that uses the proton flow to synthesise ATP from ADP (termed oxidative phosphorylation).
- This scheme represents the culmination of the oxidative energy extraction process. The energy from the reduced bonds of the fuels is conserved as a proton gradient which is then used in a combination reaction to do work on a system represented by the equilibrium between ATP, ADP and Pi (see Chapter 2, note 4).

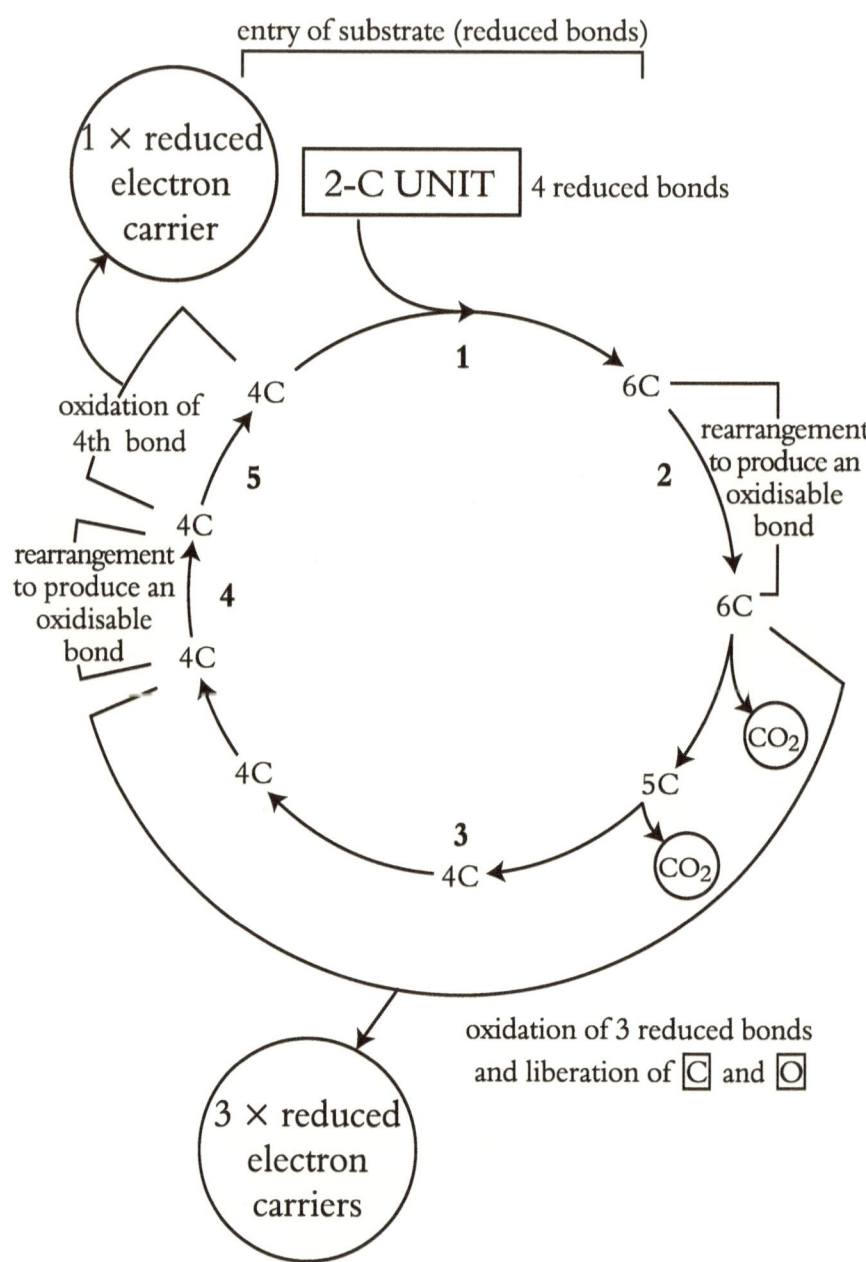

Fig. 4.3

Figure 4.3   Partitioning the Krebs cycle into functional units

- The Krebs cycle can be partitioned into five segments.
- Segment 1 is the entry segment comprising the condensation of the 2-carbon acetyl group (representing four reduced bonds and carried by CoA) with a 4-carbon member of the cycle.
- Segment 2 is a rearrangement segment, in which an oxidisable bond is produced.
- Segment 3 is where three of the four reduced bonds are oxidised, and where the carbon of the fuels exits the process as CO_2.
- Segment 4 is another rearrangement producing an oxidisable bond.
- Segment 5 is where the final (fourth) oxidation takes place and where the 4-carbon compound required in segment 1 is resynthesised.

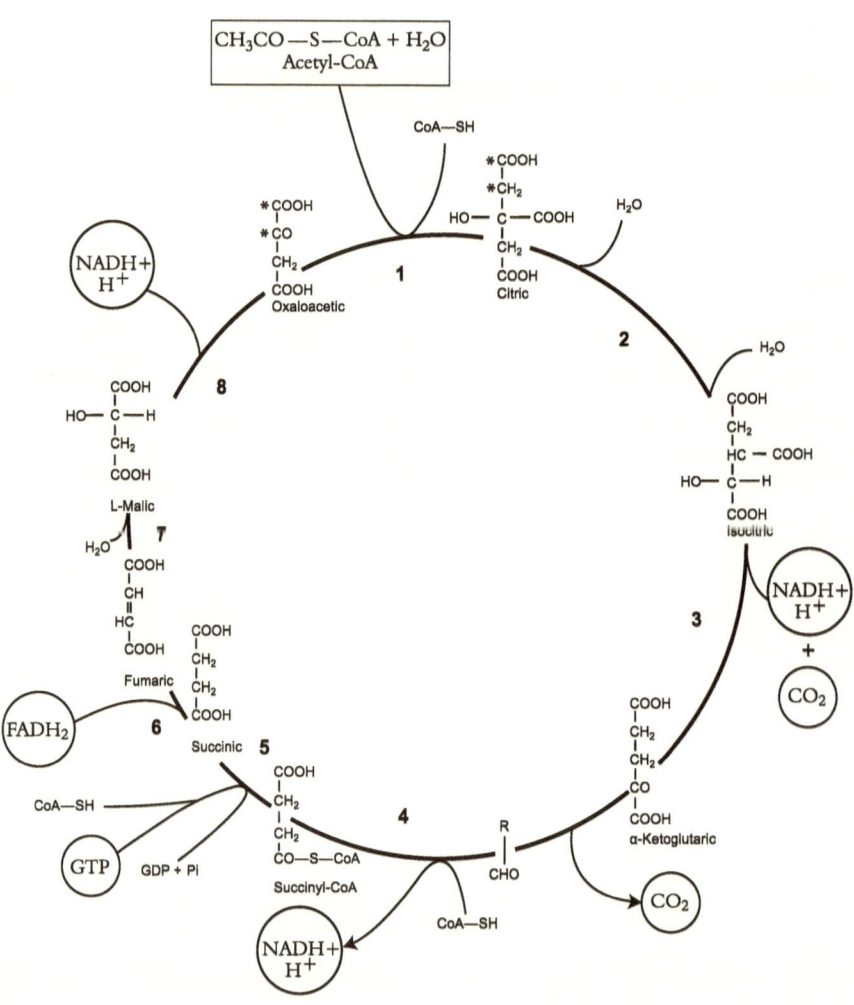

Fig. 4.4

Figure 4.4 The detailed reactions of the Krebs cycle

- Use this figure to identify the themes that are reiterated in the Krebs cycle, and which are discussed in the text.
- The reactions are numbered so the text can be related to the figure.
- The two carbons from the acetyl group of acetyl CoA are marked with an asterisk (*) in citrate and oxaloacetate.
- Adapted from Stryer, L, *Biochemistry*, fourth edition, New York, W H Freeman and Co., 1995.

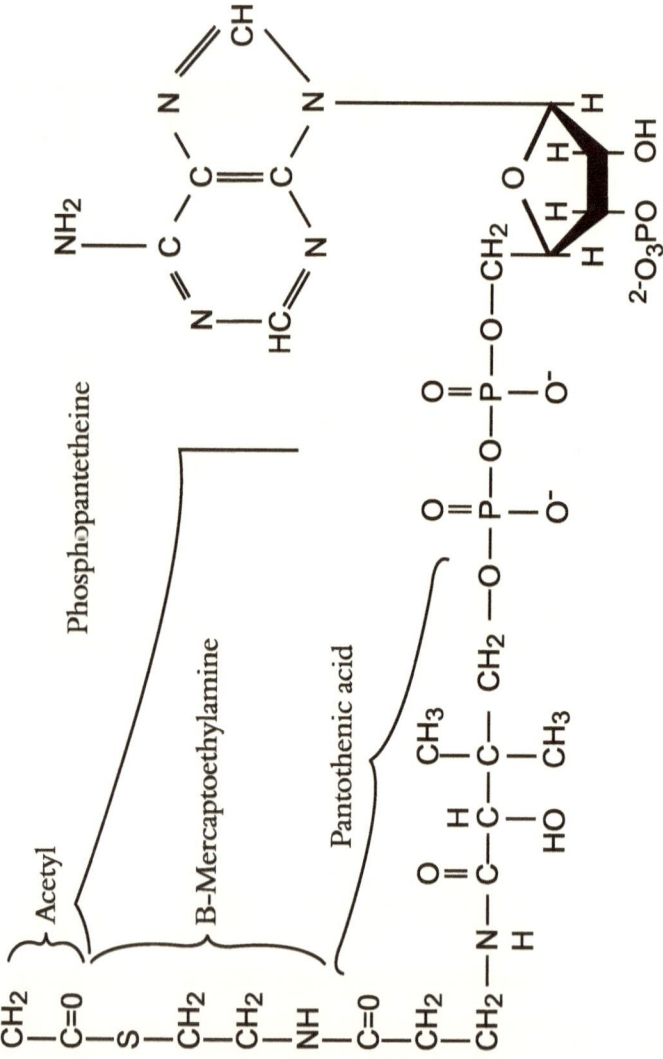

Fig. 4.5

Figure 4.5   The structure of acetyl CoA

- Note the size of the compound, but the small business-end.
- Note that AMP is part of the structure.
- CoA lacks the acetyl group and is terminated by an SH group.
- Adapted from Stryer, L, *Biochemistry*, fourth edition, New York, W H Freeman and Co., 1995.

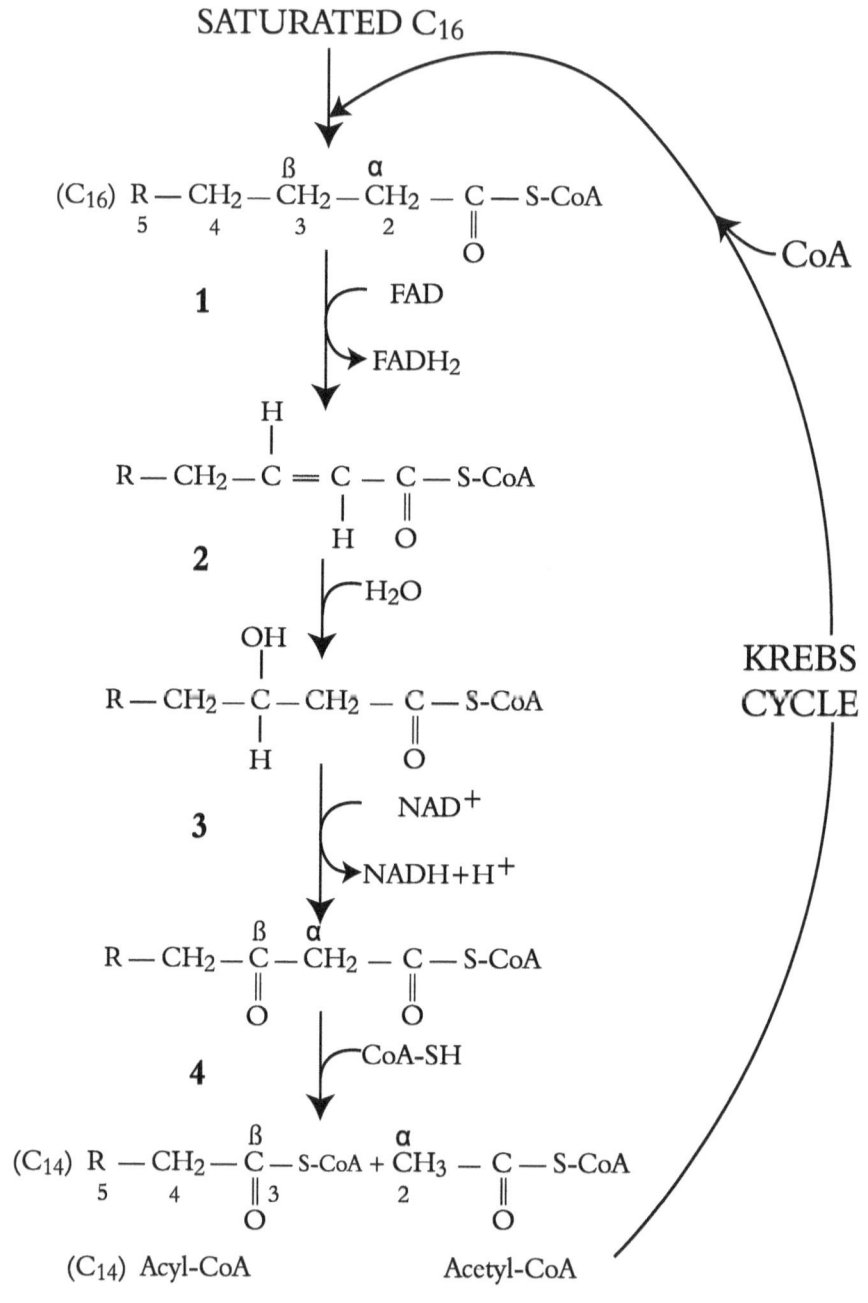

Fig. 4.6

Figure 4.6   The detailed reactions of β-oxidation

- The starting point is a long-chain acyl CoA.
- Three reactions, two oxidations and a rearrangement accomplish the oxidation of the β carbon (labelled).
- A 2-carbon acetyl CoA is split off in the final reaction as a reduced bond substrate for the Krebs cycle.
- Two reduced bonds are oxidised with each 'cycle' to produce reduced electron carriers.
- The carbons are numbered in the initial and final intermediates so carbon number can be related to the β carbon.
- Adapted from Lehninger, A L, *Principles of Biochemistry*, New York, Worth Publishers, 1982.

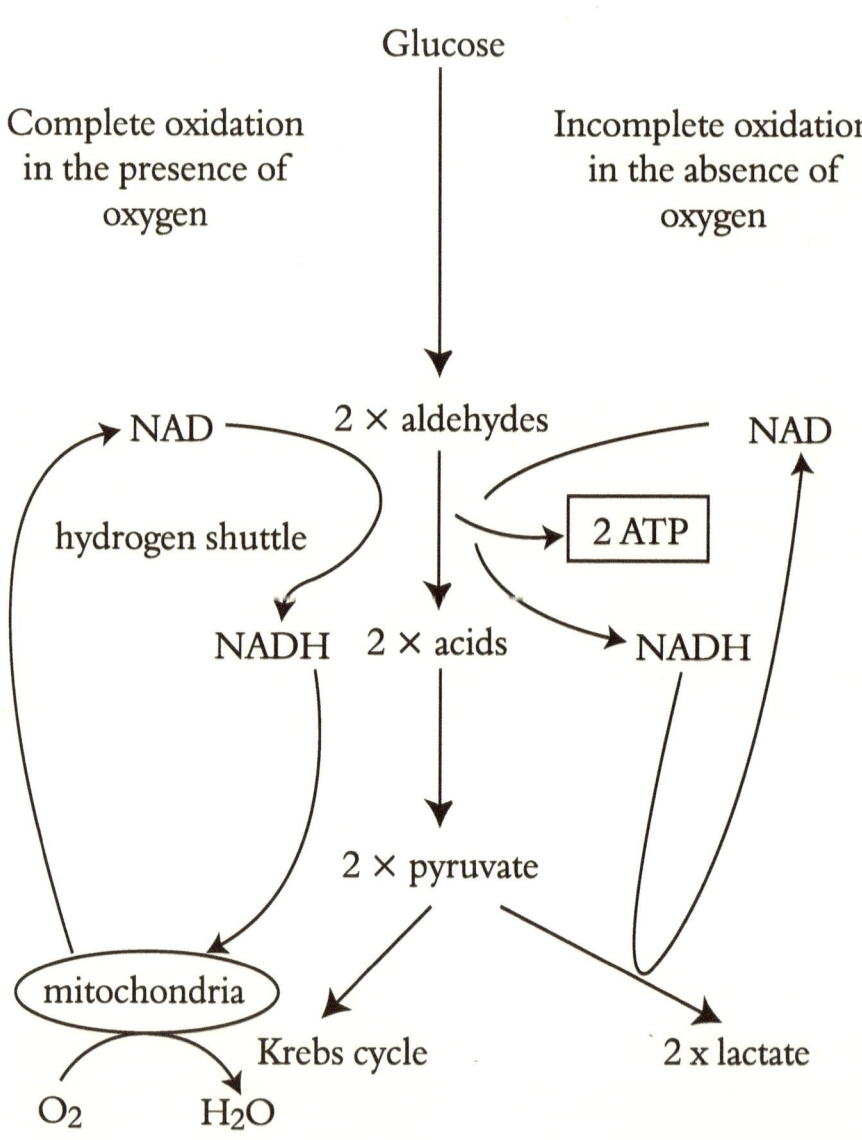

Fig. 4.7

Figure 4.7 The essentials of glycolysis

- The 6-carbon glucose is converted to two 3-carbon aldehydes.
- The aldehydes are oxidised to acids producing $NADH + H^+$ and the 'extra energy' from the aldehyde oxidation (see reaction 4 of the Krebs cycle) is conserved as ATP (concept: energetics of an aldehyde oxidation). This is the only ATP that can be produced in the absence of oxygen.
- The acids are further processed to pyruvate which is converted to acetyl CoA (in the presence of oxygen) or lactate (in the absence of oxygen). Note that in the absence of oxygen, end-products other than lactate are possible (see Chapter 3), but the concepts and themes involved remain the same.
- The $NADH + H^+$ is recycled (see repeating theme: recycling) by a hydrogen shuttle (in the presence of oxygen), or by the reaction that converts pyruvate to lactate (in the absence of oxygen).
- The structures of the various intermediates are shown in Figure 4.9.

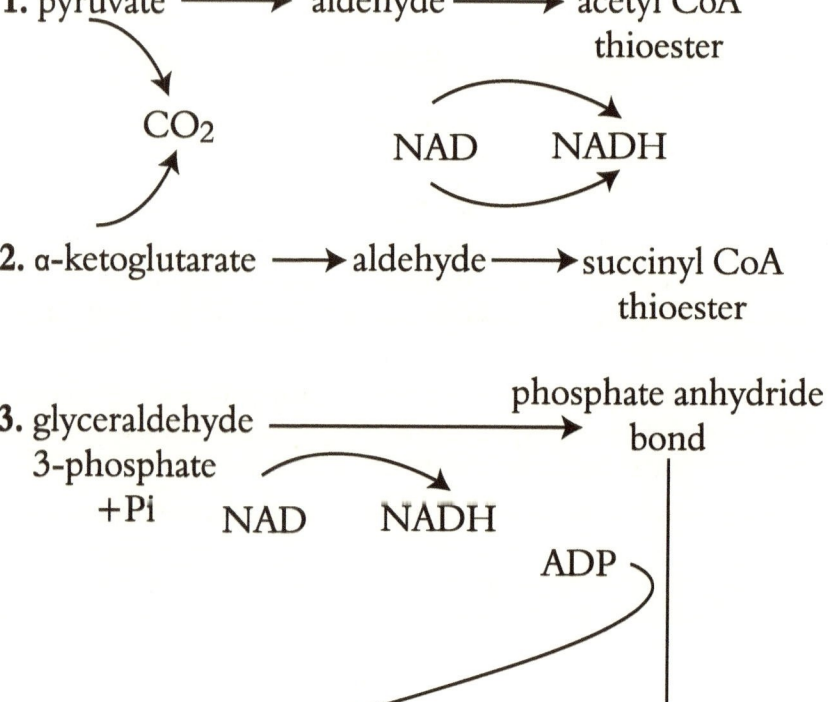

Fig. 4.8

Figure 4.8 The three aldehyde oxidation reactions

- Reaction 1 is the reaction converting pyruvate to acetyl CoA. It is virtually identical to reaction 4 in the Krebs cycle (reaction 2 in this figure). Note that the structure of pyruvate (Figure 4.9) is essentially that of the first three carbons of α-ketoglutarate.
- In reactions 1 and 2, decarboxylation produces aldehydes which are oxidised to acids producing $NADH + H^+$ and the 'extra energy' from the aldehyde oxidation is conserved as two different forms of a thioester.
- Reaction 3 (reactions 6 and 7 in Figure 4.9) is again very similar, except that the reaction starts with an aldehyde and the 'extra energy' from the aldehyde oxidation is conserved as a phosphate anhydride bond and subsequently as ATP.

Glucose

1 ATP → ADP

Glucose-6-phosphate

2

Fructose-6-phosphate

3 ATP → ADP

Fructose-1,6-bisphosphate

4

Glyceraldehyde-3-phosphate
+
Dihydroxyacetone phosphate

5

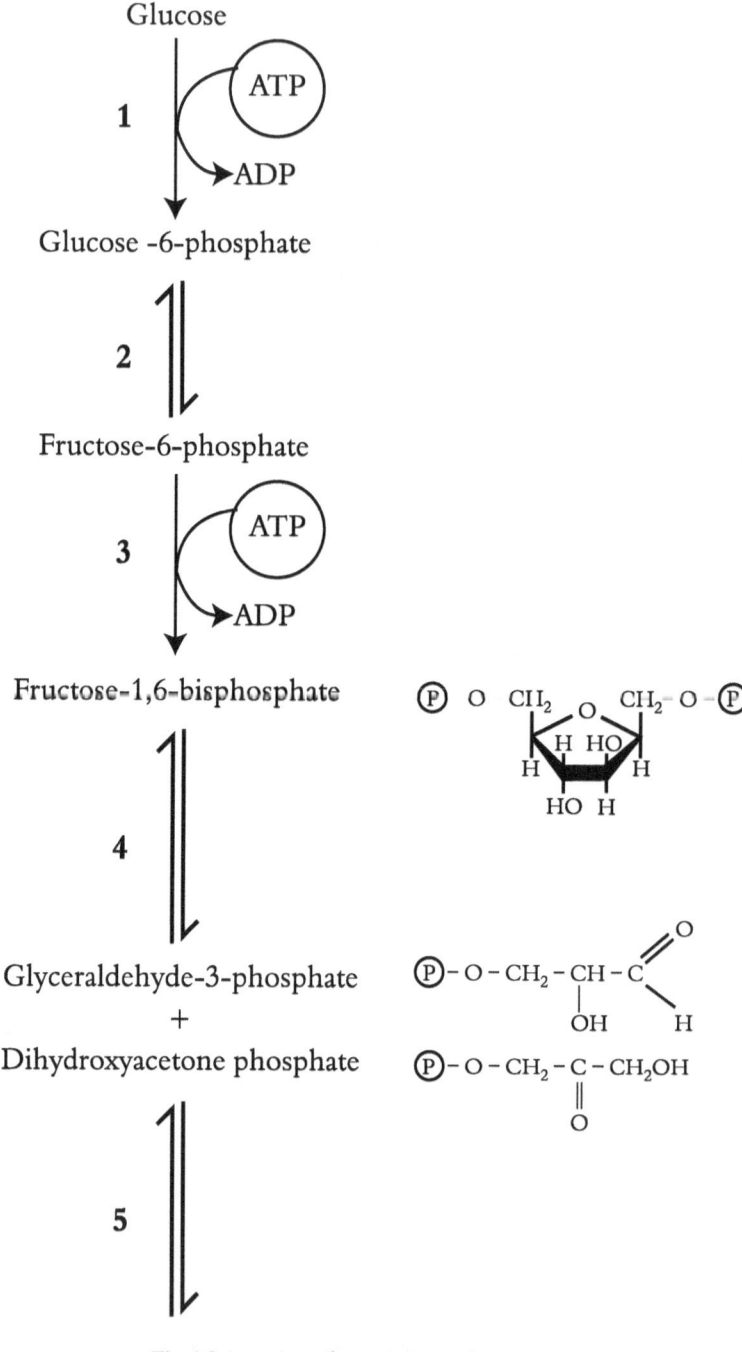

Fig.4.9 (continued on next page)

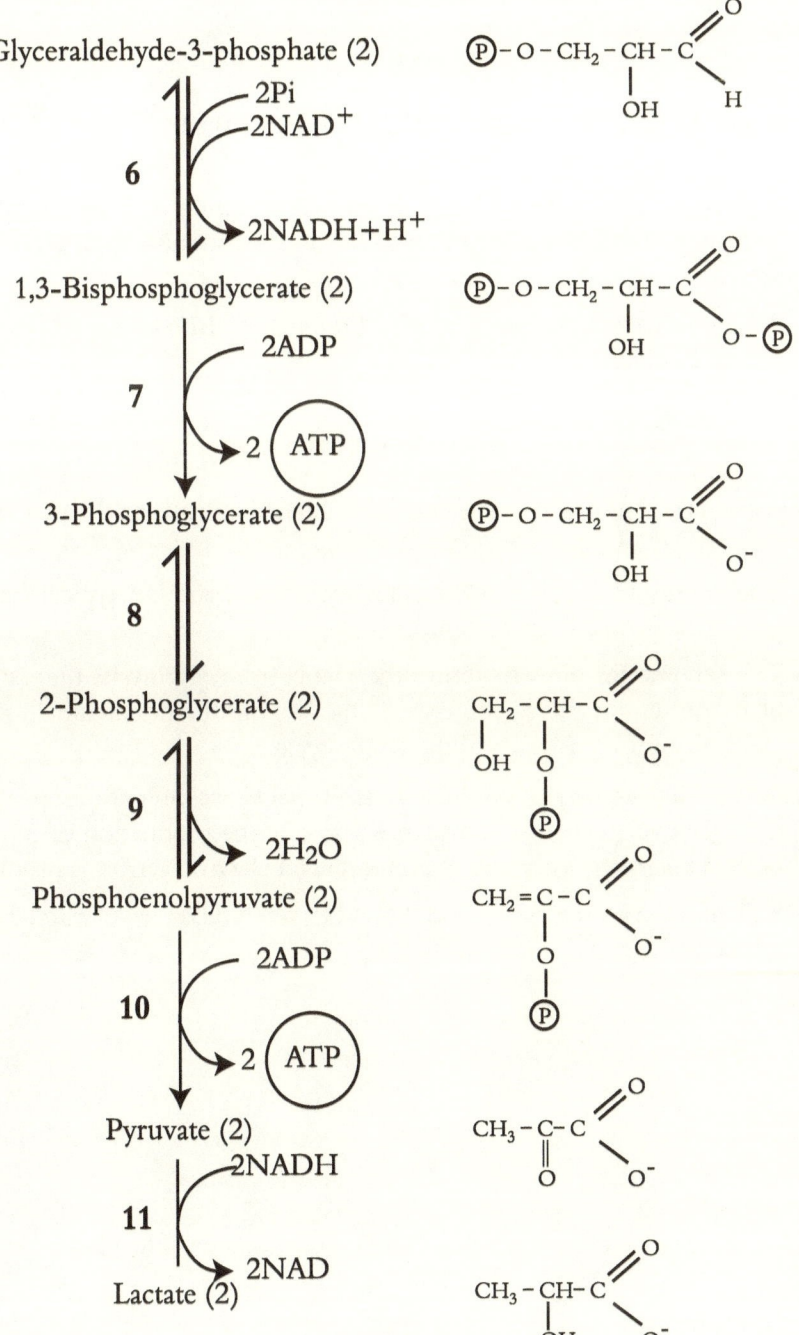

Fig.4.9 (continued)

Figure 4.9 Glycolysis in detail (previous page)

- This figure shows the structure of the relevant intermediates of glycolysis and can be used as a companion to the detailed discussion of glycolysis in the text.
- Adapted from Lehninger, et al., *Principles of Biochemistry*, New York, Worth Publishers, 1993.

Figure 4.10 The energy yield of the carbohydrate and lipid fuels

- The energy yield for two 6-carbon molecules, one is a typical carbohydrate, the other a saturated lipid.
- The scheme for glucose shows the yields from glycolysis, the oxidation of pyruvate and the Krebs cycle. The scheme for hexanoate shows the yields from β-oxidation and the Krebs cycle.
- One point that may be confusing; the initial processing of hexanoate only involves two 'cycles' of β-oxidation, despite the production of 3 × acetyl CoA. This is why only two of each reduced electron carrier is produced.
- Stoichiometry used is 2.5 and 1.5 ATP per NADH + H^+ and $FADH_2$ respectively.

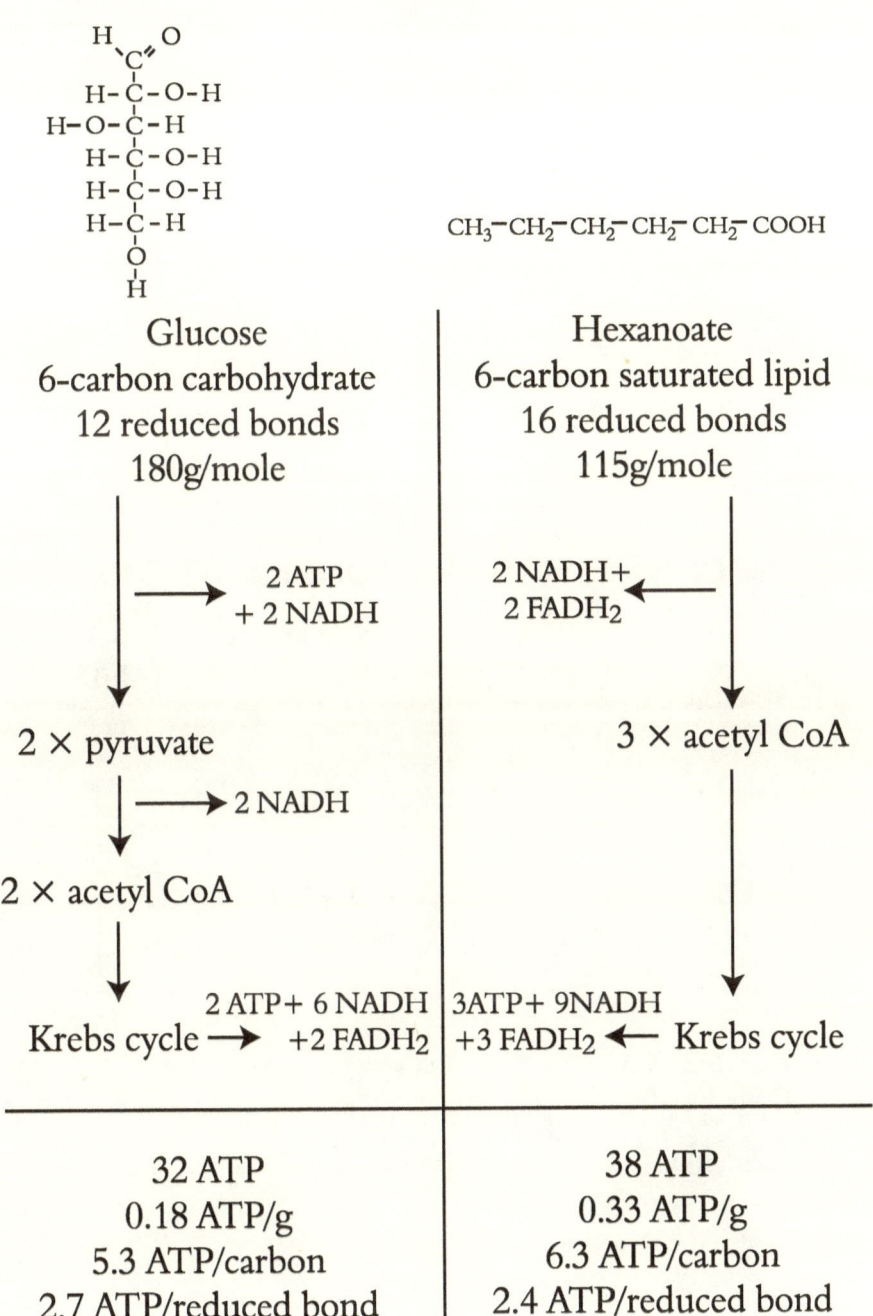

Fig. 4.10

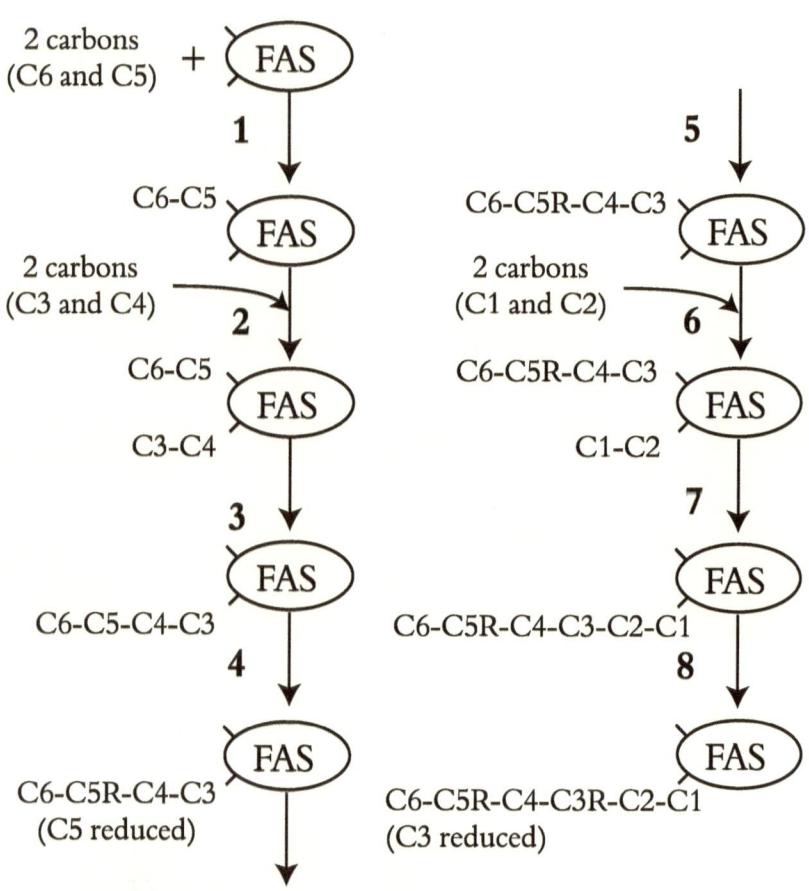

Fig. 5.1

Figure 5.1 A simplified scheme of fatty acid synthesis: the synthesis of a 6-carbon saturated fatty acid

- FAS = Fatty acid synthase complex.
- The complex contains two binding proteins that are depicted as arms sticking out of the complex.
- The reactions are numbered next to the arrows.
- The process begins with the entry and binding of a 2-carbon acetyl group. These carbons will ultimately become carbons 5 and 6 in the finished product (reaction 1).
- Following reaction 1, the pattern is simple and repetitive.
  1. Entry of the next two carbons which are bound by another binding protein (reaction 2). From this point on, this is how the carbons enter the process, reaction 1 only happens once each time a fatty acid is synthesised.
  2. Condensation (synthesis of a C-C bond; reaction 3).
  3. Reduction (synthesis of C-H bonds; reaction 4). The 'R' signifies that this carbon has been reduced.

  Note that there is not much reduction as the 2-carbon entry molecules are already mostly reduced.
- The process is then repeated as carbons 3–6 are transferred back to the original binding protein (reaction 5), C1 and 2 enter (reaction 6), another condensation occurs (reaction 7) and finally another reduction (of carbon 3; reaction 8) to produce the saturated 6-carbon fatty acid.

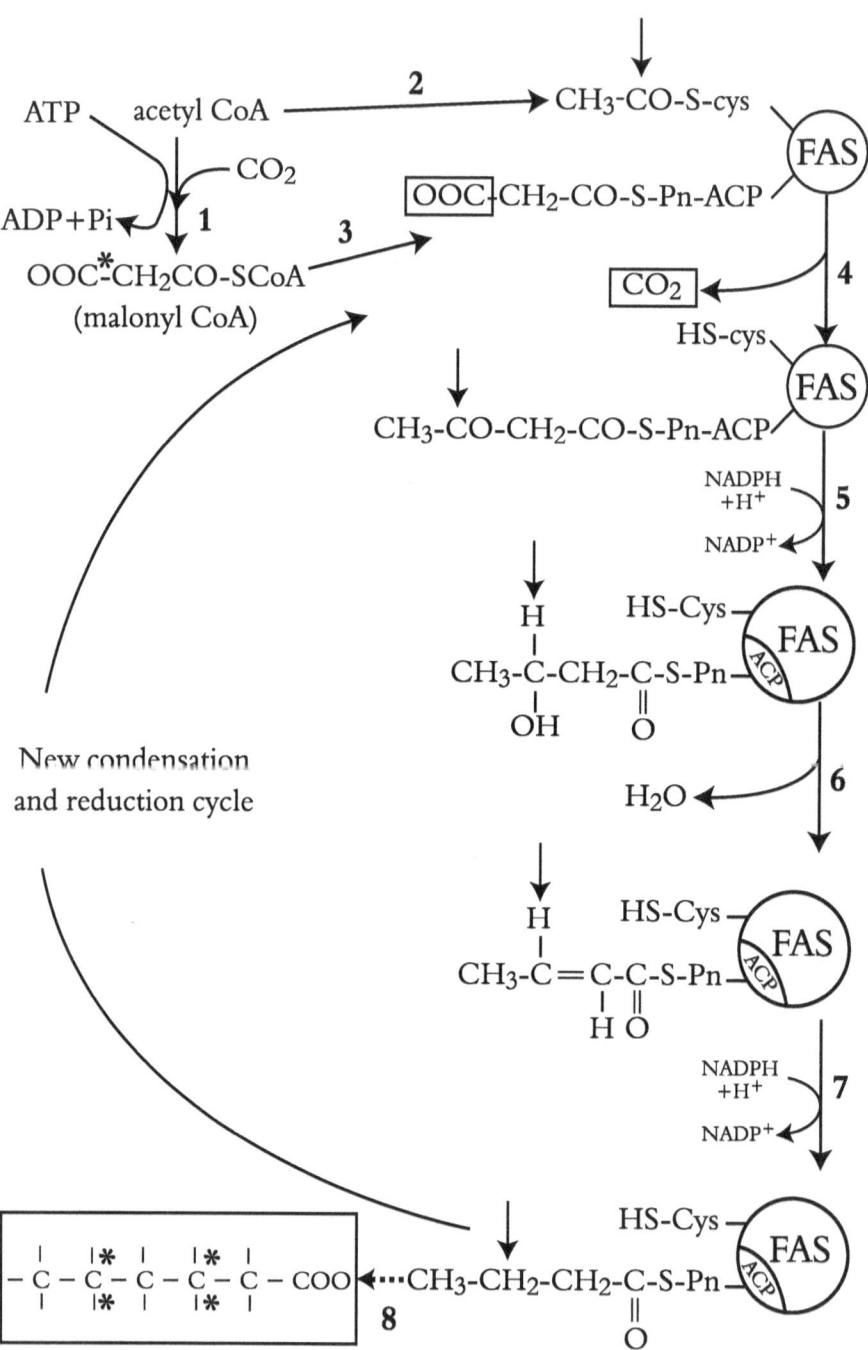

Fig. 5.2

Figure 5.2 Fatty acid synthesis: recurring concepts and themes

- Each reaction (numbered) is discussed in the text.
- The carbon that is reduced in the process is labelled with an arrow.
- cys; the cysteine residue of the condensing enzyme on FAS, which is one of the binding proteins mentioned in Figure 5.1.
- Pn; the pantothenic acid group which is bound to a protein (acyl carrier protein; ACP) that is a dedicated carrier of acyl groups.
- The reduced bonds that are synthesised during the process are marked (∗) on the ultimate 6-carbon product (boxed).
- Adapted from Lehninger, A L, Principles of Biochemistry, New York, Worth Publishers, 1982.

Figure 6.1 The synthesis of reduced bonds in plants: the Calvin cycle (overleaf)

A

- Reactions 1 to 8 describe the formation of glucose from CO_2 and ribulose 1,5-diphosphate.
- Reactions 9 to 15 describe the regeneration of the ribulose 1,5-diphosphate which is required for another cycle.
- If the reactions are carefully analysed, it can be seen that everything does actually add up. For example, the five moles of dihydroxyacetone phosphate synthesised in reaction 4, are used in reaction 5 (three of them) and in reaction 10 (two of them).
- The sum of all the reactions is shown at the end of the list of reactions. This reaction shows clearly that one mole of glucose (twelve moles of reduced bonds) is synthesised from $6CO_2$ (the carbon input) and $12\ NADPH + H^+$ (the reduced bond input), and that all other reactants and products are recycled.

B

- A more detailed version of reaction 1 (Figure 6.1.A) which is discussed in the text.
- Reproduced from Lehninger, A L, Principles of Biochemistry, New York, Worth Publishers, 1982.

$6CO_2$ + 6 ribulose 1,5-diphosphate + $6H_2O \longrightarrow$ 12 3-phosphogycerate (1)

12 3-phosphoglycerate + 12ATP $\longrightarrow$ 12 1,3 bisphosphoglycerate + 12ADP (2)

12 1,3 bisphosphoglycerate + 12NADPH + $12H^+ \longrightarrow$
$\qquad$ 12 glyceraldehyde 3-phosphate + $12NADP^+$ + 12Pi (3)

5 glyceraldehyde-3-phosphate $\longrightarrow$ 5 dihydroxyacetone phosphate (4)

3 glyceraldehyde-3-phosphate + 3 dihydroxyacetone phosphate $\longrightarrow$
$\qquad$ 3 fructose 1,6-diphosphate (5)

3 fructose 1,6-diphosphate + $3H_2O \longrightarrow$ 3 fructose 6-phosphate + 3Pi (6)

fructose 6-phosphate $\longrightarrow$ glucose-6-phosphate (7)

glucose-6-phosphate + $H_2O \longrightarrow$ $\boxed{\text{glucose}}$ + Pi (8)

---

2 fructose 6-phosphate + 2 glyceraldehyde 3-phosphate $\longrightarrow$
$\qquad$ 2 xylulose 5-phosphate + 2 erythrose 4-phosphate (9)

2 erythrose 4-phosphate + 2 dihydroxyacetone phosphate $\longrightarrow$
$\qquad$ 2 sedoheptulose 1,7-diphosphate (10)

2 sedoheptulose 1,7-diphosphate + $2H_2O \longrightarrow$
$\qquad$ 2 sedoheptulose 7-phosphate + 2Pi (11)

2 sedoheptulose 7-phosphate + 2 glyceraldehyde 3-phosphate $\longrightarrow$
$\qquad$ 2 ribose 5-phosphate + 2 xylulose 5-phosphate (12)

2 ribose 5-phosphate $\longrightarrow$ 2 ribulose 5-phosphate (13)

4 xylulose 5-phosphate $\longrightarrow$ 4 ribulose 5-phosphate (14)

6 ribulose 5-phosphate + 6ATP $\longrightarrow$ 6 ribulose 1,5-diphosphate + 6ADP (15)

$$\boxed{\text{Sum: } 6CO_2 + 18ATP + 12H_2O + 12NADPH + 12H^+ \longrightarrow C_6H_{12}O_6 + 18Pi + 18ADP + 12NADP^+}$$

Fig. 6.1.A

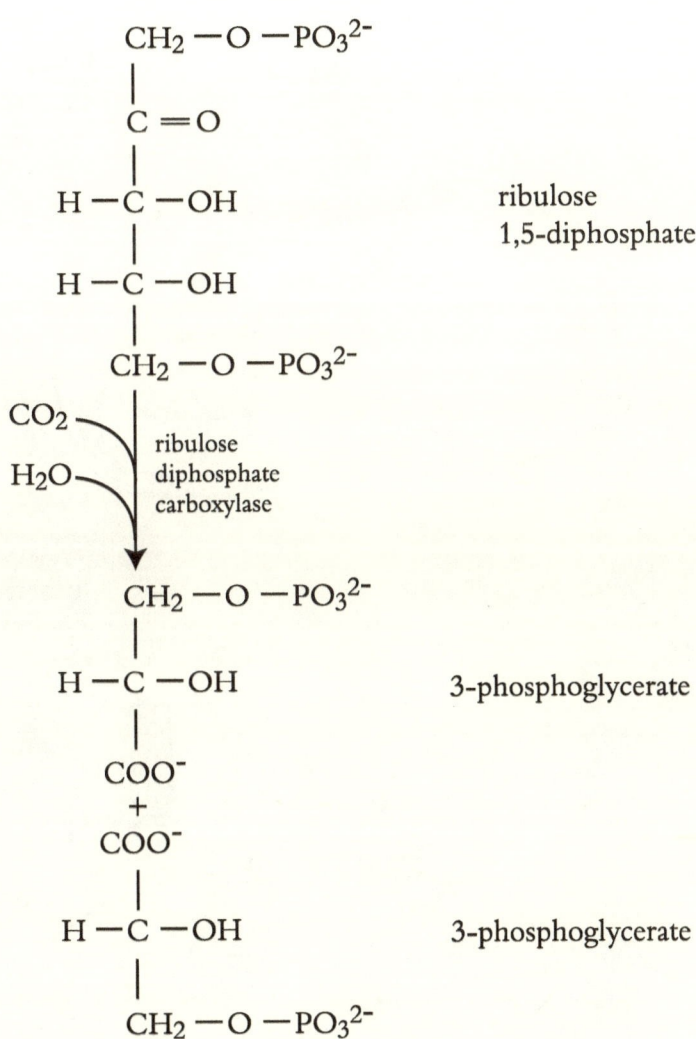

Fig. 6.1.B

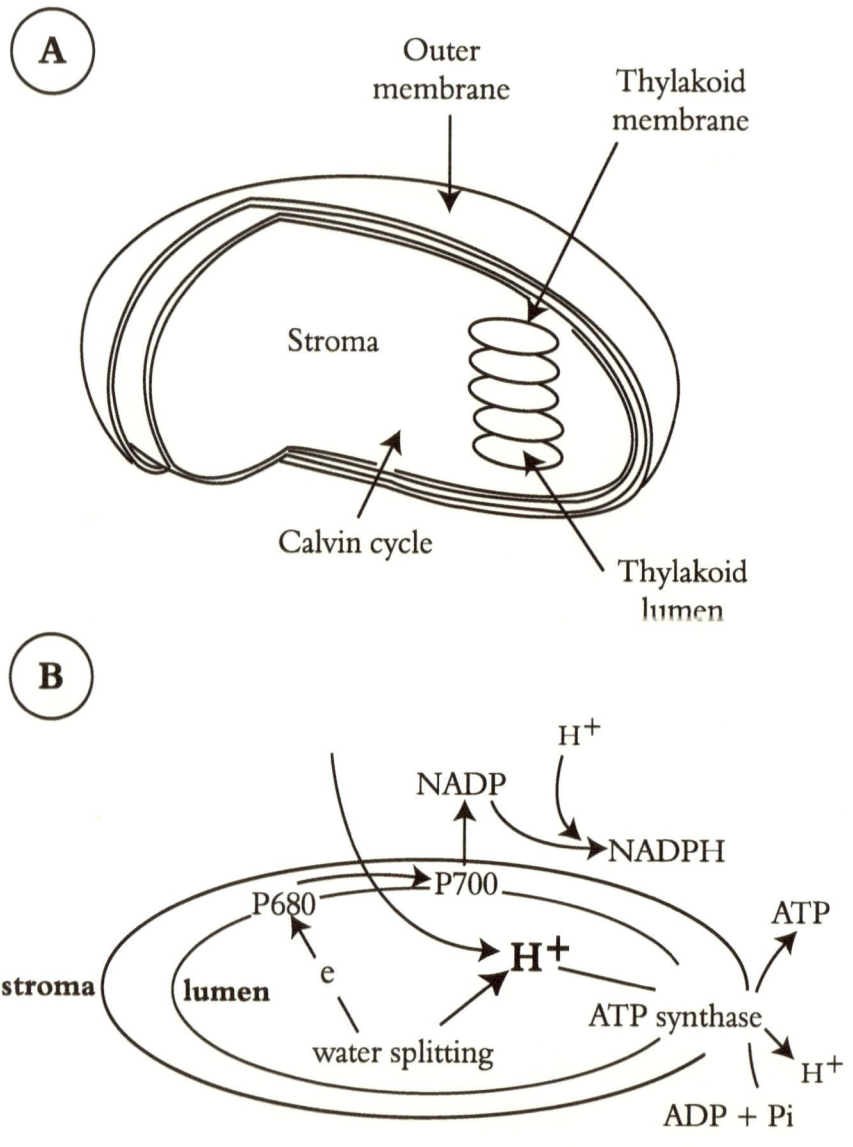

Fig.6.2

Figure 6.2 The chloroplast and the location of the elements of photo-reduction

A. THE CHLOROPLAST

- The chloroplast has an outer and inner membrane, like the mitochondria.
- The electron accepting and donating complexes, as in the mitochondria, are located in a membrane, but in this case in the membrane of the thylakoid invaginations.

B. THE THYLAKOID

- Water is split in the lumen of the thylakoid.
- The electrons are donated to P680 (see below), then to P700 and finally to NADP to form $NADPH + H^+$.
- The water is split, using light energy, by a water-splitting enzyme complex containing a Mn electron acceptor/donator centre.
- Both P680 and P700 are reaction centres which contain chlorophyl molecules. They are components of membrane-bound photosystems. These centres absorb light and transfer light-excited electrons to an electron acceptor in an electron transfer system. The photosystems are multisubunit, transmembrane complexes.
- During electron transfer, a proton gradient is formed across the thylakoid membrane, and ATP is synthesised as the gradient dissipates through an ATP synthase protein.
- Adapted from Stryer, L, Biochemistry, fourth edition, New York, W H Freeman and Co., 1995.

FIGURES

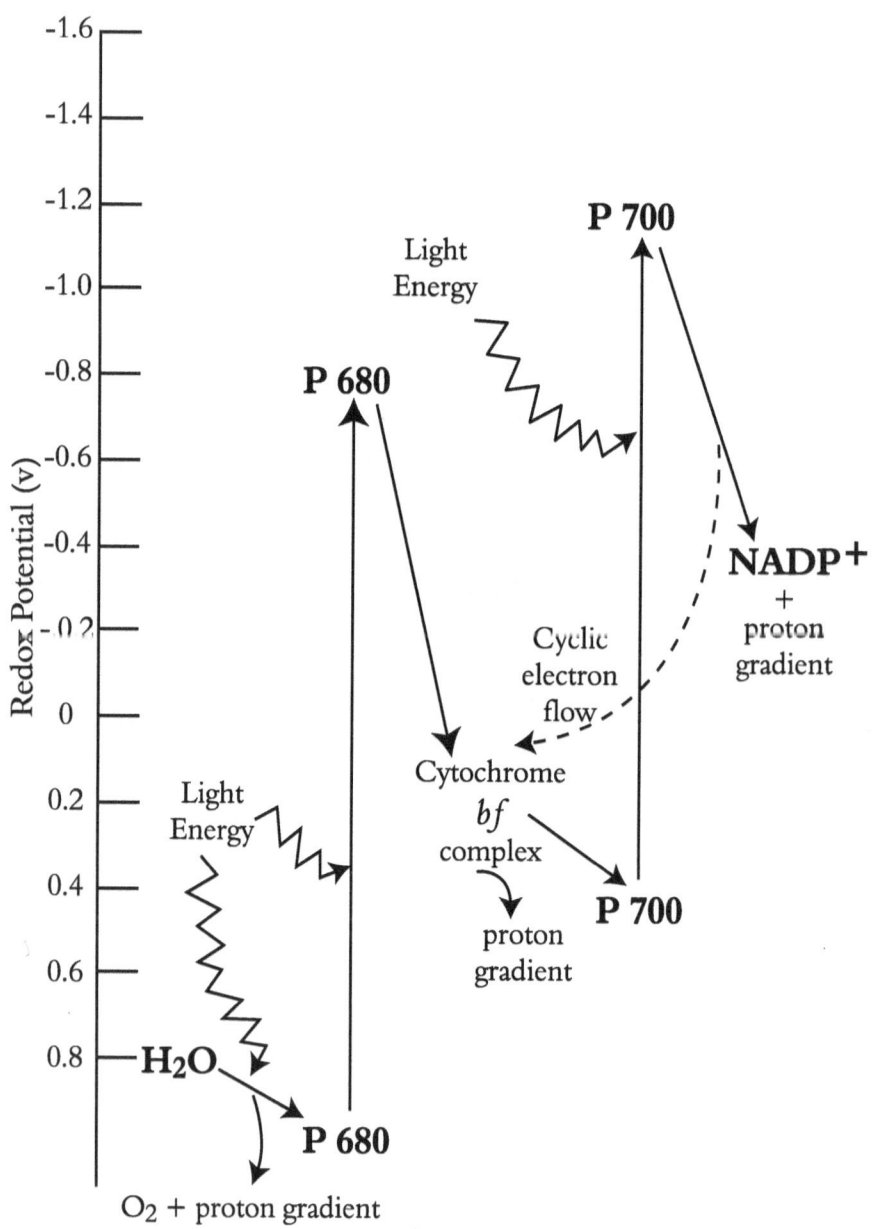

Fig.6.3

Figure 6.3 Light energy, redox potentials and the proton gradient in chloroplasts

- The scale on the left-hand side shows the redox potential (E'_0; V) of the various components depicted in the figure.
- Electrons are transferred from water to the P680 reaction centre, which has a higher affinity (higher redox potential) for electrons than the oxygen/water pair.
- The redox potential of the P680 reaction centre is then decreased through the action of light.
- Electrons are transferred from the excited P680 (E'_0 = approximately -0.8 V), through a series of acceptor/donators to another reaction centre, P700.
- The redox potential of P700 is then decreased to approximately -1.2 V, which is less than that of the NADP/NADPH + H^+ pair.
- Electrons can then be transferred from P700 to NADP, again through a series of electron acceptor/donators.
- Adapted from Stryer, L, Biochemistry, fourth edition, New York, W H Freeman and Co., 1995.

Indices

Index of Concepts

energetics of an aldehyde oxidation, the, 30, 34, 35, 46, 52, 95
energy availability from a reaction out of equilibrium, 4, 25, 52
free energy of reactions, 7
implications of high K_{eq}, 6
innocuous end-products, 22, 29, 34, 35, 52, 57, 80, 81
light energy can change redox potential, 48, 49
reactions at equilibrium and work, 3
redox potential, 13–14, 26, 37, 47, 52, 83
reduced bonds, 10, 52
using ATP hydrolysis as part of a combination reaction with a large and negative ΔG, 16, 26, 28, 30, 32, 36, 47
using ATP hydrolysis to increase the K_{eq} of a reaction, 36, 42

Index of Repeating Themes

business-ends, 14, 26, 28
NAD and FAD specificity, 15, 29, 30, 31, 38, 69
oxidation reactions, 13, 21, 29
pathway repeat, 32, 43, 46
preparation reactions, 13, 27, 29, 30, 31, 32, 33, 43
prosthetic groups, 27, 36, 42, 83
reaction repeat
    aldehyde oxidation, 34, 35
recycling, 29, 32, 34, 95
structural repeats
    AMP, 15, 16, 28
    ATP synthase, 27, 50
    phosphate anhydride bond, 36
    phosphopantetheine, 28, 32, 42
    sulphydryl groups, 36, 42

General Index

acetate, anoxic end-product, 53
acetyl CoA
    as substrate of Krebs cycle, 22, 28, 80
    carrier of reduced bonds, 19, 51, 87
    in β-oxidation, 32
    in fatty acid synthesis, 41
    structure, 91
alanopine, anoxic end-product, 53
aldehyde oxidation, 97, *See also* Index of Concepts
alkane
    oxidation in Krebs cycle, 31
AMP. *See* Index of Repeating Themes
arginine
    role in ATP production, 20
ATP
    aerobic vs anoxic production, 20–21, 33–37
    chemistry and structure of, 15–17
    end-product of fuel oxidation, 51, 61
    hydrolysis as part of a combination reaction, 16, 61, 73
    hydrolysis driving C-C bond formation, 41
    hydrolysis reaction, 16, 72
    bypassing, 47
    per reduced bond, 53
    produced in Krebs cycle, 30
    production from NADH and FADH, 37
    synthesis in electron transfer system, 25–26, 85
    synthesis in photosynthesis, 45, 49–50, 109
ATP synthase. *See also* Index of Repeating Themes
    ADP as a substrate, 85
    in chloroplasts, 50, 109
    in mitochondria, 26, 85
β-oxidation
    as a preparation pathway, 19
    details of, 31, 93
    energy extraction by, 21, 79
    functions, 21
    summary, 79
bonds
    C-C and C-H, 8–10, 12, 25, 40–41, 51, 63, 65, 103
    C-C conversion to C-H, 29, 30, 42
    C-C synthesis is not net reduction, 49
    C-C, not oxidised, 13, 29, 33
    C-H
        electrons removed from, 13
        oxidation of, 13, 33
    C-N and N-H, 11
    C-O, 10
    oxidised in fuels, 63, 67
    phosphate anhydride, 36, 47, 72, 97, *See also* Index of Repeating Themes
    reduced. *See* reduced bonds
        oxidation of, 64
        synthesised in photosynthesis, 45, 49
    with oxygen, 9
brain
    glucose, use of, 34
    lactate production by, 53
business-ends. *See* Index of Repeating Themes
Calvin cycle, 45–47
    carbon input, 46
    synthesis of reduced bonds in, 46
carbohydrate synthesis, 44
carbon dioxide
    form in which C and O exit the oxidation process, 19, 87

product of oxidation reactions, 9, 12, 51, 61, 65
chemical reactions
  at equilibrium, 3
  out of equilibrium, 4, 6, 17, 72, *See also* Index of Concepts
  oxidation, three types of, 13, 21, 43, 67, *See also* Index of Repeating Themes
  preparation and rearrangement, 13, 51, *See also* Index of Repeating Themes
  proceeding towards equilibrium, 3, 25–26, 51, 61
  reduction, 40, 51
  role of redox potentials in electron transfer, 71
  used by cells to do work, 4
  work done by, 3–4
chloroplast, 48–50, 51, 109
citric acid cycle. *See* Krebs cycle
CO_2. *See* carbon dioxide
combination reactions. *See* Index of Concepts
electron acceptors other than oxygen, 53–54
electron carriers
  ATP production from NADH and FADH, 37
  FMN, 27, 69, 83
  NAD and FAD, 13, 69
  NAD, recycling, 34
  NADP, 43, 69
  photosynthesis, 48–50, 109, 111
  redox potentials of, 13, 70
  reduced
    produced by glycolysis and β-oxidation, 19
    produced in Krebs cycle, 22, 27, 80
  role in electron transfer system, 23, 25
  role in oxidation of reduced bonds, 13–15, 51
  role in reduction, 40
  specificity of NAD and FAD, 44, 69, 71, *See also* Index of Repeating Themes
electron transfer system in animals
  ADP as a substrate for, 81
  details of, 25, 83
  exit of hydrogen from, 81
  functions, 23–24
  substrates and products of, 23–24
  summary, 81
electron transfer system in plants
  details, 48–50, 109, 111
electrons
  affinity for, 10, 11
  donating to oxygen to produce energy, 15, 51
end-products
  innocuous, 22, 80, 81, *See also* Index of Concepts
  other than CO_2 and H_2O, 52
energy
  available to cells, 1, 3, 8–10
  conserved as proton gradient, 39, 50
  consumed at rest and during exercise in humans, 10, 55, 56
  consumed during marathon, 10
  content of cellular fuels, 10, 37–38, 55–56, 100
  extra energy. *See* aldehyde oxidation
  extracted by electron transfer system, 23
  extracted by the Krebs cycle, 22
  extraction by β-oxidation, 21
  extraction by glycolysis, 21
  extraction by oxidation, 65
  from nuclear fission, 8–9
  from nuclear power stations, 8
  from nuclear weapons, 8
  in a reduced bond, 10
  light. *See* light energy
  nucleus, 8
  useable, 15
ethanol

as a product of anoxic ATP
   production, 21, 53, 77
extra energy. *See* aldehyde oxidation
fatty acid synthase, 40
fatty acid synthesis
   acetyl CoA in, 41
   cysteine in, 105
   details, 41, 105
   formation of C-C bonds in, 41, 103
   formation of C-H bonds in, 40–41, 103
   malonyl CoA in, 42
   pantothenic acid in, 105
   reduction pathway, as, 40, 44
   simplified scheme, 103
fatty acids
   ATP production from without oxygen, 21
   efficient fuel, 57
   reduced bonds of, fate of, 32
fire vs biological oxidation, 19
free energy
   actual and redox reactions, 14
   actual, of a reaction, 7–8
   conserved as ATP, 16
   of a reaction, 7
   of formation of molecules, 3, 30
   of reactions
      negative standard, implications, 28, *See also* Index of Concepts
   standard, of a reaction, 7–8
   standard, of ATP hydrolysis, 17
   total, of a solution, 3
fuels
   classes of, 8, 10, 62
   energy in cellular, 8, 37, 100
   long-chain fatty acid, 21
   oxidation of, 9, 12, 61, 63, 64, 75
   reduced, 1, 51
   reduced bonds in, 62, 63
   reduction state, lipids vs carbohydrate, 38
   reduction using solar energy, 1, 65
   releasing energy from, 9

used by hummingbirds and insects, 10
glucose
   conversion to lipid, 57
   reduced bonds in, 45, 49, 62, 63
   synthesis in photosynthesis, 45–47, 49
glycolysis
   as a preparation pathway, 20–21
   details, 33, 95, 100
   energy extraction by, 21
   functions, 20, 33–35
   inhibition during sprinting, 35
   production of ATP without oxygen, 21, 35
   summary, 77
GTP, 30, 39
H_2O. *See* water
hydrogen
   exit from oxidation process as water, 51, 81
   shuttle, 34, 35, 95
   transport across mitochondrial membrane, 25
innocuous end-product. *See* end-products, innocuous
K_{eq}
   for oxidation of glucose, 9
   implications of high, 6, 28, 36, 42, 43, *See also* Index of Concepts
   implications of low, 47
   increased by involving ATP. *See* Index of Concepts
   of ATP hydrolysis, 17, 72
Krebs cycle
   details, 27, 87, 89
   functions, 22
   oxidation of reduced bonds in, 80
   summary of, 80
lactate
   advantages as an end-product, 52
   end-product of glycolysis, 21, 35, 77
   produced during 100 m sprint, 21, 52
   production by parasitic worms, 53

reduced bonds, 35
light energy
  effect on redox potentials, 48, *See also* Index of Concepts
  role in photosynthesis, 48–49
malonyl CoA, 42
marathon
  fuel constraints, 35
metabolic pathways. *See also* Index of Repeating Themes
  variations on standard themes, 20
metabolic rate at rest and during exercise in humans, 10, 55, 56
muscle
  contraction without oxygen, 35
  fuels for during marathon, 35
oxidation
  biological vs fire, 19
  of carbohydrate and lipids, 19, 75
  partial, by parasitic worms, 53
  products, 19
  reactants, 19
  three types of reaction, 13, 21, 29, 67
oxygen
  electrons, affinity for, 9–10, 11, 23, 26
  energy production, role in, 23
  link between oxidation and reduction, 49, 51
  redox potential, 23
  substrate in oxidation process, 23, 61, 63, 65, 81, 83
pantothenic acid. *See* phosphopantetheine
pentose phosphate pathway, 33
phosphopantetheine. *See* Index of Repeating Themes
photosynthesis
  ATP synthesis in, 45, 49
  carbon input, 105
  glucose synthesised in, 45
  reduced bonds synthesised in, 45–46, 51, 105
  role of light energy, 45, 48, 111
  role of NADP, 45, 46, 47–49, 105

role of water, 47–50, 109
prosthetic groups. *See also* Index of Repeating Themes
  in fatty acid synthase, 40
  in the electron transfer system, 83
proton gradient
  energy conserved as, 25–26, 39, 50, 85
  in animal electron transfer system, 39, 83
  in plant electron transfer system, 50, 109, 111
pyruvate
  fates of, 20, 77
  product of glycolysis, 20
  reduced bonds of, 34
  reduction in glycolysis, 35
rearrangement of molecules. *See* chemical reactions
recycling. *See* Index of Repeating Themes
redox potential. *See also* Index of Concepts
  in animal electron transfer system, role of, 26, 27, 83
  in photosynthesis, role of, 47, 48–49, 111
  of electron carriers, 13, 70
  of NADP, 43
  of oxygen, 23, 70
reduced bonds. *See also* Index of Concepts
  concept of, 2, 8–9, 62
  conversion to ATP, 26
  from non-organic sources, 54
  numbers of in a molecule, 67, 71
  oxidation in Krebs cycle, 27, 80, 87
  per cent synthesised in animals, 44, 51
  ranking in terms of energy production, 38
  sources of for cell, 19
  synthesis of, 38, 51, 105
solvent capacity of a cell, 6, 11
  related to reactions in photosynthesis, 47

sulphydryl groups. *See* Index of Repeating Themes
thioester bond, hydrolysis of, 28
thylakoid, 48, 109
tricarboxylic acid cycle. *See* Krebs cycle
water
   end point of oxidation, 51, 81
   form in which H exits the oxidation process, 20, 81
   link between oxidation and reduction, 49
   photosynthesis, role in, 47–50, 109, 111
   product of oxidation reactions, 9, 12, 23, 61, 65
weight loss
   implications of lipid vs carbohydrate, 10
   role of exercise and diet, 55–57
   role of resting metabolic rate, 56
   strategy in humans, 55–57

www.ingramcontent.com/pod-product-compliance
Lightning Source LLC
Chambersburg PA
CBHW031922240526
45464CB00022B/637